石油高等院校特色规划教材

高等油藏工程

（双语版）

张继成　刘　丽　孙丽艳　编

石油工业出版社

内 容 提 要

本书以复杂油气藏的高效开发为基础,以建立最优化开发系统为主线,从油田开发核心原理、驱油效率计算方法、波及系数及其计算方法、油田动态分析的经验方法和油田开发调整理论与技术等几个方面,进行了系统、深入和前瞻性的介绍与解读,可为读者提供系统的高等油藏工程基础知识及相关的英文表达方式。

本书可作为石油高等院校石油与天然气工程相关专业研究生双语教材,也可供油田企业和研究院所的其他相关专业人员作为参考书和培训教材使用。

图书在版编目(CIP)数据

高等油藏工程:汉、英 / 张继成,刘丽,孙丽艳编.
—北京:石油工业出版社,2023.11
石油高等院校特色规划教材
ISBN 978 - 7 - 5183 - 6036 - 9

Ⅰ.①高… Ⅱ.①张… ②刘… ③孙… Ⅲ.①油藏工程—高等学校—教材—汉、英 Ⅳ.①TE34

中国国家版本馆 CIP 数据核字(2023)第 098581 号

出版发行:石油工业出版社
(北京市朝阳区安华里二区 1 号楼　100011)
网　　址:www.petropub.com
编辑部:(010)64256990
图书营销中心:(010)64523633　(010)64523731
经　　销:全国新华书店
排　　版:北京密东文创科技有限公司
印　　刷:北京中石油彩色印刷有限责任公司

2023 年 11 月第 1 版　2023 年 11 月第 1 次印刷
787 毫米×1092 毫米　开本:1/16　印张:16.5
字数:335 千字

定价:49.00 元
(如发现印装质量问题,我社图书营销中心负责调换)
版权所有,翻印必究

前　　言

  油藏工程是一门从总体上认识评价和指导改造油气藏的技术学科,是一门高度综合的学科,涉及地球物理、石油地质、油层物理、渗流力学及采油工程等方面的原理、方法和成果资料。高等油藏工程是针对各种类型的复杂油气藏,为建立或完善高效开发系统,结合适宜和先进的开发技术而进行油藏工程设计理论和方法研究并提供油藏开发设计方案和调整方案的总称。高等油藏工程是油藏工程基础的具体深化及应用,其主线依然是建立最优化开发系统,但是目标油藏更具复杂性、特殊性和个性化特征,除静态及动态监测资料解释、物质平衡方程和经验方法外,驱油效率、波及系数及以此为基础的开发指标计算也成为高等油藏工程动态分析和效果预测的重要内容。

  "高等油藏工程"是东北石油大学为石油与天然气工程各方向的硕士研究生设置的专业必修课程,课程采用双语教学。本书分为两部分,第一部分为英文部分,第二部分为中文部分,每一部分分为5章,可帮助读者在学习专业知识的同时掌握相关的英文表达。为了保持英文部分和中文部分各自的特色及表达的相对独立性,两部分内容不完全对应。

  本书由东北石油大学石油工程学院张继成、刘丽和孙丽艳共同编写完成,具体编写分工如下:张继成负责中文篇第一章、第三章、第五章和英文篇第一章、第三章、第五章的编写;刘丽负责中文篇第四章和英文篇第四章的编写;孙丽艳负责中文篇第二章和英文篇第二章的编写。在编写过程中,参考了国内外学者的相关文献,在此致以由衷的谢意!

  由于编者水平有限,书中难免存在不足和不当之处,敬请读者提出宝贵意见。

编者

2023 年 3 月

目 录

英 文 篇

Chapter 1　Key Principles on Oilfield Development Engineering ……………… 3
　Section 1　Energy-Resistance Theory ……………………………………………… 3
　Section 2　Arrangement of Well Pattern …………………………………………… 6
　Section 3　Effect of Trapped Gas on Waterflooding Recovery ………………… 10
　Section 4　Grouping of Development Zones ……………………………………… 15
　Section 5　Economic Evaluation …………………………………………………… 20
　Section 6　Oilfield Development Plan ……………………………………………… 23

Chapter 2　Method to Calculate Displacement Efficiency …………………… 26
　Section 1　Definition of Recovery Factor ………………………………………… 26
　Section 2　Fractional Flow Equation ……………………………………………… 28
　Section 3　Frontal Advance Equation ……………………………………………… 35
　Section 4　Oil Recovery Calculations ……………………………………………… 45

Chapter 3　Method to Calculate Sweep Efficiency …………………………… 51
　Section 1　Areal Sweep Efficiency ………………………………………………… 51
　Section 2　Vertical Sweep Efficiency ……………………………………………… 78
　Section 3　Method to Predict Recovery Performance for Layered Reservoir ……… 85

Chapter 4　Empirical Methods for Dynamic Analysis ………………………… 91
　Section 1　Oil Production Decline Law …………………………………………… 91
　Section 2　Rising Law of Water Cut ……………………………………………… 99
　Section 3　Joint Solution of Type B Water Drive Law Curve and Weibull Model ……… 110

Chapter 5　Theory and Technology for Oilfield Development Adjustment ……… 120
　Section 1　Monitoring, Analysis and Evaluation of Informational Data in Oilfield
　　　　　　Development ……………………………………………………………… 120
　Section 2　Content of Oilfield Development Adjustment ………………………… 125

Section 3　Experience and Prospect of Development Adjustment in Lamadian, Saertu, and Xingshugang Oilfields ·· 128

Section 4　Discussion on Revolutionary Technology for Oilfield Development ··············· 140

中　文　篇

第一章　油田开发核心原理 ·· 147
 第一节　能耗原理 ·· 147
 第二节　井网部署 ·· 149
 第三节　气体滞留对水驱采收率的影响 ·· 152
 第四节　开发层系 ·· 156
 第五节　经济评价 ·· 160
 第六节　油田开发方案 ·· 163

第二章　驱油效率计算方法 ·· 164
 第一节　采出程度定义 ·· 164
 第二节　分流方程 ·· 165
 第三节　前缘推进方程 ·· 172
 第四节　生产动态指标计算 ··· 181

第三章　波及系数及其计算方法 ··· 186
 第一节　平面波及系数 ·· 186
 第二节　纵向波及系数 ·· 208
 第三节　多层油藏开发动态预测方法 ·· 214

第四章　油田动态分析的经验方法 ·· 219
 第一节　油田产量递减规律 ··· 219
 第二节　含水上升规律 ·· 225
 第三节　乙型水驱规律曲线与威布尔预测模型的联解法 ··· 234

第五章　油田开发调整理论与技术 ·· 244
 第一节　油田开发信息监测与分析评价 ··· 244
 第二节　油田开发调整的内容 ·· 248
 第三节　喇萨杏油田开发调整经验与展望 ··· 250
 第四节　关于变革性开采技术的探讨 ·· 256

参考文献 ·· 258

英文篇

Chapter 1 Key Principles on Oilfield Development Engineering

Section 1 Energy-Resistance Theory

Oilfield development is a process in which fluid consumes energy, overcomes resistance, generates movement, and flows from the formation to the ground. The whole process of oilfield development is related to energy and resistance, and the effect of oilfield development also depends on energy and resistance.

1. Development Mechanism

Development mechanism refers to that what kind of energy initial oil in place depends on to flow from formation to bottom of oil well. The development mechanism corresponds to flow event in porous medium, which refers to the way to provide energy in flow event, including natural energy and artificially supplemented energy.

1.1 Natural energy

Six development mechanisms basically provide the natural energy necessary for oil recovery: (1) Rock and liquid expansion. (2) Solution gas drive. (3) Gas cap drive. (4) Natural water drive. (5) Gravity drainage drive. (6) Combination drive.

1.2 Artificial energy supplement

If natural energy is not enough to support fluid to overcome resistance and flow from formation to bottom of oil well, human intervention is needed to supplement energy to formation fluid through water injection and other artificial means. Water injection is to drill a well to inject treated water that meets quality standard into formation. The injected water carries energy into pore-throat network in formation, contacts crude oil, and delivers energy to crude oil. Crude oil gets energy supplement, it continues to overcome resistance and flow to bottom of production well.

2. Efficiency of Energy

Energy is the key factor dominating the effect of oilfield development. Two aspects about it are crucial: energy magnitude and energy efficiency. In order to obtain excellent oilfield development results, on the one hand, we should maintain high level of magnitude in energy, on the other hand, we should ensure high energy efficiency. For oil fields with high water cut and ultra-high water cut that have been developed for a long time, it may be easy to maintain the level of energy, and the most important work is to find ways to improve the efficiency of energy.

According to Darcy's law, the relationship between liquid production rate and energy meets the following equation:

$$Q_\mathrm{L} = \frac{\text{Energy}}{\text{Resistance}} \qquad (1-1)$$

The higher the energy magnitude is, the greater the liquid production rate.

The relationship between oil production rate and energy meets the following formula:

$$Q_\mathrm{O} = \frac{\text{Effective energy}}{\text{Resistance}} = \frac{\alpha \times \text{Energy}}{\text{Resistance}} \qquad (1-2)$$

Oil production rate is related not only to energy level, but also to energy efficiency. Maintaining energy level may not guarantee oil production rate. This is especially true for oilfields with high and ultra-high water cut.

As shown in Figure 1-1, the multiple layer reservoir system is composed of three oil layers with different permeability, using the same injector and producer for water flooding.

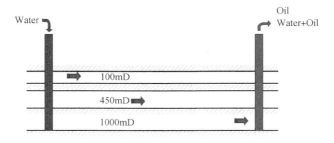

Figure 1-1 Schematic diagram of a multiple layer system

The injected water advances rapidly along the channels with good permeability and water front reaches the production well first along these channels. After water breakthrough at oil well, the water saturation of these channels increases and the flow resistance decreases, so the water volume allocated into these channels and the water advancing speed along these channels will increase, the water saturation of these channels will continue to increase and the flow resistance will further decrease with the increase of water saturation and water proceeding speed, which in turn will increase the water saturation and water propulsion speed along these channels. With this cycling continues, these channels will eventually turn into inefficient circulation channels.

There are two reasons for the formation of "inefficient circulation of injected water". One is the heterogeneity of geological factors, such as permeability, thickness and fault, etc. The second is the heterogeneity of development conditions, such as the heterogeneity of injection-production well spacing, and operating behavior of oil and water wells, etc.

3. Technical Ways to Improve Energy Efficiency

The technical way to improve energy efficiency is mainly to block off the inefficient channel and increase its resistance to liquid through technologies such as profile control and water plugging. Then the subsequent injected water will be diverted to the channel with low resistance and more remaining oil. As a result, energy efficiency and development effect are enhanced.

Although profile control and water shut-off are widely used measures, the technology of them is mature and the technical effect is also good, but there are also serious shortcomings. For example, some recoverable reserves are lost. As shown in Figure 1 − 2, the pore-throat network is a three-dimensional body composed of highly complex pores and throats. The injected plugging agent blocks the inefficient channels. However, due to the complicated and heterogeneous connection of the inefficient channels and non inefficient channels, a part of the pore-throat network enriched in remaining oil will be surrounded and blocked by the injected plugging agent, the reserves inside these porous media will be more difficult to tap in the future, or even completely lost.

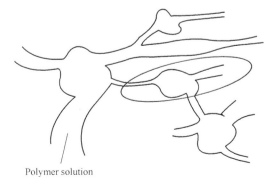

Figure 1 − 2　Schematic diagram of plugging in pore-throat network

In addition, the measures of profile control and water shut-off also mean that more energy is expended. Plugging the inefficient channel with less remaining oil, high water saturation and low resistance will inevitably increase the resistance of the whole system, and will inevitably consume more energy when the liquid production remains unchanged.

Profile control and water shut-off measures can significantly improve energy efficiency, and the effect is very good. In the process of practical application, we need to pay attention to the above two problems and make continuous improvement in order to enhance the level of oilfield development to a much greater extent.

Section 2　Arrangement of Well Pattern

Injection pattern refers to the locations of oil production wells and water injection wells and their arranging mode in a water flooding oilfield, i. e. , the arrangement of well pattern.

Essentially four types of well arrangements are used in fluid injection projects: irregular injection patterns, peripheral injection patterns, regular injection patterns, crestal and basal injection patterns.

1. Irregular Injection Patterns

Willhite (1986) points out that surface or subsurface topology and/or the use of slant-hole drilling techniques may result in production or injection wells that are not uniformly located. In these situations, the region affected by the injection well could be different for every injection well. Some small reservoirs are developed for primary production with a limited number of wells and when the economics are marginal, perhaps only few production wells are converted into injectors in a nonuniform pattern. Faulting and localized variations in porosity or permeability may also lead to irregular patterns.

2. Peripheral Injection Patterns

In peripheral flooding, the injection wells are located at the external boundary of the reservoir and the oil is displaced toward the interior of the reservoir (Figure 1 - 3). Craig (1971), in an excellent review of the peripheral flood, points out the following main characteristics of the flood:

(1) The peripheral flood generally yields a maximum oil recovery with a minimum of produced water.

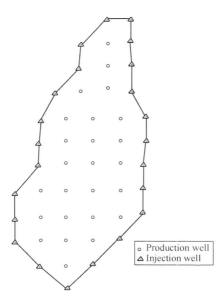

Figure 1-3 Typical peripheral water flood

(2) The production of significant quantities of water can be delayed until only the last row of producer remains.

(3) Because of the unusually small number of injectors compared with the number of producers, it takes a long time for the injected water to fill up the reservoir gas space. The result is a delay in the field response to the flood.

(4) For a successful peripheral flood, the formation permeability must be large enough to permit the movement of the injected water at the desired rate over the distance of several well spacings from injection wells to the last line of producers.

(5) To keep injection wells as close as possible to the waterflood front without by passing any movable oil, watered-out producers may be converted into injectors. However, moving the location of injection wells frequently requires laying longer surface water lines and adding costs.

(6) Results from peripheral flooding are more difficult to predict. The displacing fluid tends to displace the oil bank past the inside producers, which are thus difficult to produce.

(7) Injection rates are generally a problem because the injection wells continue to push the water greater distances.

3. Regular Injection Patterns

Due to the fact that oil leases are divided into square miles and quarter square miles, fields are developed in a very regular pattern.

Direct line drive (Figure 1-4). The lines of injection wells and production wells are directly opposed to each other. The pattern is characterized by two parameters: a = distance between wells of the same type, and d = distance between lines of injectors and producers.

Staggered line drive (Figure 1-5). The wells are in lines as in the direct line, but the injectors and producers are no longer directly opposed but laterally displaced by a distance of $a/2$.

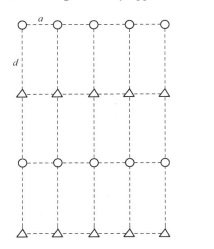
Figure 1-4 Direct line drive

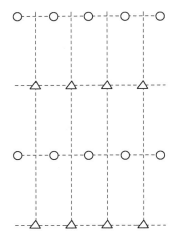
Figure 1-5 Staggered line drive

Five spot (Figure 1-6). This is a special case of the staggered line drive in which the distance between all like wells is constant, i. e., $a = 2d$. Any four injection wells thus form a square with a production well at the center.

Seven spot (Figure 1-7). The injection wells are located at the corner of a hexagon with a production well at its center.

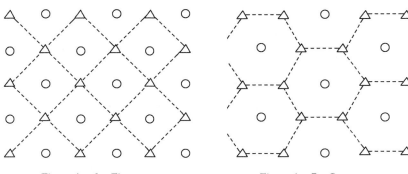
Figure 1-6 Five spot   Figure 1-7 Seven spot

Nine spot (Figure 1-8、Figure 1-9). This pattern is similar to that of the five spot but with an extra injection well drilled at the middle of each side of the square. The pattern essentially contains eight injectors surrounding one producer.

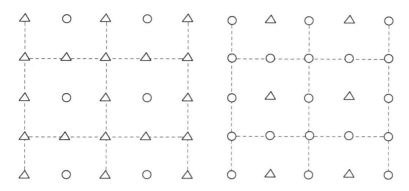

Figure 1 – 8 Normal nine spot   Figure 1 – 9 Inverted nine spot

The patterns termed inverted have only one injection well per pattern. This is the difference between normal and inverted well arrangements. Note that the four spot and inverted seven spot patterns are identical (Figure 1 – 10、Figure 1 – 11).

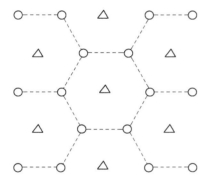
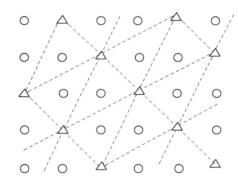

Figure 1 – 10 Four spot (Inverted seven spot)   Figure 1 – 11 Skewed four spot

4. Crestal and Basal Injection Patterns

In crestal injection, as the name implies, the injection is through wells located at the top of the structure. Gas injection projects typically use a crestal injection pattern. In basal injection, the fluid is injected at the bottom of the structure. Many water-injection projects use basal injection patterns with additional benefits being gained from gravity segregation (Figure 1 – 12).

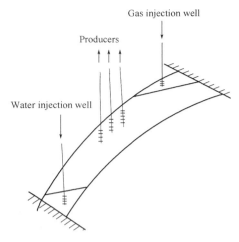

Figure 1 – 12 Well arrangements for dipping reservoirs

Section 3　Effect of Trapped Gas on Waterflooding Recovery

Numerous experimental and field studies have been conducted to study the effect of the presence of initial gas saturation on waterflood recovery. Early research indicated that the waterflooding of a linear system results in the formation of an oil bank, or zone of increased oil saturation, ahead of the injection water. The moving oil bank will displace a portion of the free gas ahead of it, trapping the rest as a residual gas (Figure 1 – 13). Several authors have shown through experiments that oil recovery by water is improved as a result of the establishment of trapped gas saturation, S_{gt}, in the reservoir.

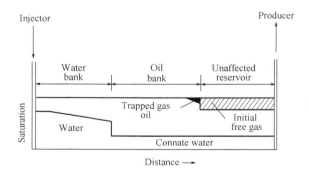

Figure 1 – 13 Water saturation profile during a waterflood

The theory of this phenomenon of improving overall oil recovery when initial gas exists at the start of the flood is not well established; however, Cole (1969) proposed the following two different theories that perhaps provide insight to this phenomenon.

1. First Theory

Cole (1969) postulates that since the interfacial tension of a gas oil system is less than the interfacial tension of a gas-water system, in a three-phase system containing gas, water, and oil, the reservoir fluids will tend to arrange themselves in a minimum energy relationship. In this case, this would dictate that the gas molecules enclose themselves in an oil "blanket." This increases the effective size of any oil globules, which have enclosed some gas. When the oil is displaced by water, the oil globules are reduced to some size dictated by the flow mechanics. If a gas bubble existed on the inside of the oil globule, the amount of residual oil left in the reservoir would be reduced by the size of the gas bubble within the oil globule.

As illustrated in Figure 1 - 14, the external diameters of the residual oil globules are the same in both views. However, in view b, the center of the residual oil globule is not oil, but gas. Therefore, in view b, the actual residual oil saturation is reduced by the size of the gas bubble within the oil globule.

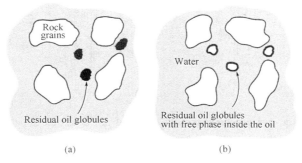

Figure 1 - 14 Effect of free gas saturation on S_{or} (first theory)

2. Second Theory

Cole (1969) points out that reports on other laboratory experiments have noted the increased recovery obtained by flooding cores with air after waterflooding.

These cores were classified as water-wet at the time the laboratory experiments were conducted.

On the basis of these experiments, it was postulated that the residual oil saturation was located in the larger pore spaces, since the water would be preferentially pulled into the smaller pore spaces by capillary action in the water-wet sandstone.

At a later time, when air was flooded through the core, it moved preferentially through the larger pore spaces since it was nonwetting.

However, in passing through these large pore spaces, the air displaced some of the residual oil left by water flooding.

This latter theory is more nearly compatible with fluid flow observations, because the gas saturation does not have to exist inside the oil phase.

If this theory were correct, the increased recovery due to the presence of free gas saturation could be explained quite simply for water-wet porous media.

As the gas saturation formed, it displaced oil from the larger pore spaces, because it is more nonwetting to the reservoir rock than the oil.

This phenomenon is illustrated in Figure 1 – 15. In view a, there is no free gas saturation and the residual oil occupies the larger pore spaces. In view (b), free gas saturation is present and this free gas now occupies a portion of the space originally occupied by the oil. The combined residual saturations of oil and gas in view (b) are approximately equal to the residual oil saturation of view a.

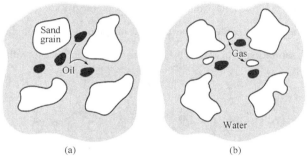

Figure 1 – 15 Effect of free gas saturation on S_{or} (second theory)

Craig(1971) presented two graphical correlations that are designed to account for the reduction in the residual oil saturation due to the presence of the trapped gas (Figure 1 – 16、Figure 1 – 17).

$$S_{gt} = a_1 + a_2 S_{gi} + a_3 S_{gi}^2 + a_4 S_{gi}^3 + \frac{a_5}{S_{gi}}$$

$$\Delta S_{or} = a_1 + a_2 S_{gt} + a_3 S_{gt}^2 + a_5 S_{gt}^3 + \frac{a_5}{S_{gt}}$$

$a_1 = 0.030517211$
$a_2 = 0.4764700$
$a_3 = 0.69469046$
$a_4 = -1.8994762$
$a_5 = -4.1603083 \times 10^{-4}$

Figure 1 – 16 Relation between S_{gt} and S_{gi}

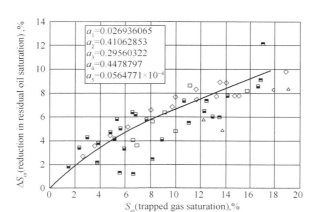

Figure 1-17 Effect of S_{gt} on waterflood recovery

Khelil suggests that waterflood recovery can possibly be improved if a so-called "optimum gas saturation" is present at the start of the flood.

This optimum gas saturation is given by:

$$(S_g)_{opt} = \frac{0.001867 K^{0.634} B_o^{0.902}}{\left(\dfrac{S_o}{\mu_o}\right)^{0.352} \left(\dfrac{S_{wi}}{\mu_w}\right)^{0.166} \phi^{1.152}} \quad (1-3)$$

where $(S_g)_{opt}$——optimum gas saturation;

S_o, S_{wi}——oil and initial water saturations;

μ_o, μ_w——oil and water viscosities, cP;

K——absolute permeability, mD;

B_o——oil formation volume factor;

ϕ——porosity.

The above correlation is not explicit and must be used in conjunction with the material balance equation (MBE). The proposed methodology of determining $(S_g)_{opt}$ is based on calculating the gas saturation as a function of reservoir pressure (or time) by using both the MBE and the above equation. When the gas saturation as calculated by the two equations is identical, this gas saturation is identified as $(S_g)_{opt}$.

The injection into a solution gas drive reservoir usually occurs at injection rates that cause repressurization of the reservoir.

If pressure is high enough, the trapped gas will dissolve in the oil with no effect on subsequent residual oil saturations.

It is of interest to estimate what pressure increases would be required in order to dissolve the trapped gas in the oil system.

The pressure is essentially defined as the "new" bubble-point.

As the pressure increases to the new bubble-point pressure, the trapped gas will dissolve in the oil phase with a subsequent increase in the gas solubility from R_s, to R_s^{new}.

As illustrated in Figure 1-18, the new gas solubility can be estimated as the sum of the volumes of the dissolved gas and the trapped gas in the reservoir divided by the volume of stock-tank oil in the reservoir,

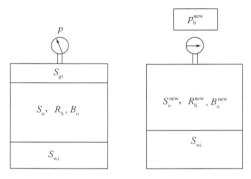

Figure 1-18 Variable bubble-point pressures

$$R_s^{new} = \dfrac{\dfrac{S_o V_p}{B_o} R_s + \dfrac{S_{gt} V_p}{B_g}}{\dfrac{S_o V_p}{B_o}} \tag{1-4}$$

$$R_s^{new} = R_s + \dfrac{S_{gt}}{S_o} \dfrac{B_o}{B_g} \tag{1-5}$$

where R_s^{new}——gas solubility at the "new" bubble-point pressure;

V_p——pore volume;

R_s——gas solubility at current pressure p;

B_g——gas formation volume factor;

B_o——oil formation volume factor;

S_{gt}——trapped gas saturation.

The pressure that corresponds to the new gas solubility (R_s^{new}) on the R_s vs. p relationship is then identified as the pressure at which the trapped gas will completely dissolve in the oil phase.

Section 4　Grouping of Development Zones

For a multiple layer reservoir, there are large difference between oil layers, so it is necessary to divide all oil layers into several groups of development zones. A group of development zones is produced with one set of well pattern alone. In someone oilfield, the number of groups of development zones can be as much as 7 or 8 sets.

1. Significance of Grouping of Development Zones

While separate zone production technologies are adopted in a certain number of wells, almost all of the world's large oilfields with multiple layers are developed by dividing groups of development zones.

1.1　Reasonably grouping the development zones can play the full role of various oil layers

Reasonably dividing oil layer into groups of development zones is a fundamental way to develop multiple layer oilfields. The so-called group of development zones is a combination of oil layers with similar characteristics, which is produced with an independent well pattern. Based on group of development zones, production planning, dynamic research and adjustment are executed.

In one oilfield, the sedimentary environment conditions and other situations of oil zones vary in vertical well interval, the characteristics of oil layers differ from each other. Therefore, the interlayer contradiction will inevitably appear during development process. If high permeability layer and low permeability layer are produced together, the production capacity of low permeability layer is often restricted due to its larger resistance. When low-pressure layer and high-pressure layer are produced together, the low-pressure layer often does not produce oil, and even that, the oil in the high-pressure layer may flow into the low-pressure layer. For a water drive oilfield, high permeability layers are often water flooded quickly, which will aggravate the contradictions between layers in the case of commingled production, and result in mutual interference between oil and water layers. This will seriously affect the recovery efficiency. As shown in Figure 1 – 19.

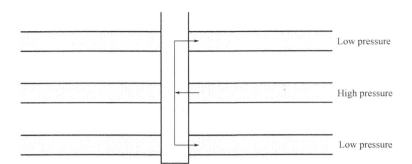

Figure 1-19 Schematic diagram of a counter flow

1.2 Grouping of development zones is the basis for deploying well patterns and planning production facilities

With groups of development zones determined, number of well patterns is fixed, and so that we can investigate and deploy wells, injection and production strategy and ground facilities. Each group of development zones in an oilfield should be independently considered for development design and adjustment. Its well pattern, injection production system, technological measures, etc., should be independently specified. As shown in Figure 1-20.

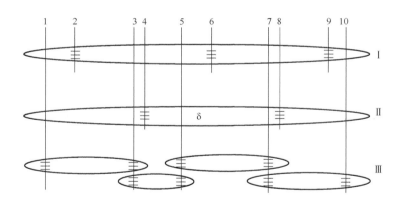

Figure 1-20 Schematic of well patterns corresponding groups of development zones

1.3 The limitation of capability of present production technologies requires dividing groups of development zones

A multiple layer oilfield often has as many as up to tens of oil-bearing zones. Producing interval in one well sometimes can be several hundred meters. The function of production technologies is to get all kinds of zones playing their roles to the maximum extent, to get them absorbing injected water uniformly and outputting crude oil uniformly. Separate zone water injection, separate zone oil production and other separate zone measures are widely applied technologies. However, due to the complexity of geological conditions, the effect of present separate zone production technologies has certain ceilings. So that, it is necessary to divide groups of development zones, ensuring the number of layers in one group of zones is not too much and the well interval corresponding it is not too long. Figure 1 – 21 shows the schematic diagram of separate zone production string.

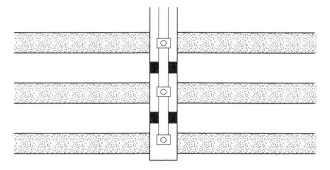

Figure 1 – 21 Schematic diagram of separate zone production string

1.4 High velocity development of oilfield requires division of groups of development zones

For a multiple layer oilfield, in order to employ all major zones to the best extent and thus to achieve a high oil output velocity, we must divide all layers into certain number of groups of development zones. This practice can play the roles of all producing layers, enhance the oil production velocity, fasten the development process, shorten the time of oilfield development process, and improve the running efficiency of fundamental investment.

2. Principles for Dividing Groups of Development Zones

Summarizing the experience and lessons on dividing groups of development zones in domestic and foreign countries' oilfields, we should take into consideration the following principles.

(1) Combine layers with similar characteristics into the same one group of development zones, to assure that all layers in the same one group have common adaptability to water injection strategy and well pattern, and that the interlayer interference is to the least extent. Similar properties are mainly reflected in similar sedimentary condition, similar permeability, similar extending area, and similar heterogeneity.

Generally, engineers take oil-bearing formation as the basic unit to combine groups of development zones. According to lots of research work and filed practice, some oilfields propose to divide and combine groups of development zones using sandstone formation as the basic unit, because one sandstone formation is an independent sedimentary unit layers inside which having similar properties.

(2) An independent group of development layers should have certain amount of crude oil reserves to ensure that it meets the designed oil production velocity, has a long time of stable production, and achieves better economic indexes.

(3) There must be good barrier bed between two neighboring groups of development zones, so that in the process of injected water flooding, different groups of development zones can be strictly separated apart and there is no communication and interference between different groups.

(4) The structure shape, oil water contact, pressure system and properties of crude oil in the same group of development zones should be relatively similar.

(5) Within the range separate zone stimulation technology can solve, not too more groups of development zones should be divided, so as to reduce the construction workload and improve economic benefits.

3. Limit of the Number of Layers in One Group of Development Zones

The degree of difference between layers in one group of development zones is used as the first consideration when determining the limit of number of layers in this group. In the other words, it is to investigate the best way of division and combination of groups that can give full play to production capacity of each layer in the same one group of development zones. Interlayer interference is a complex process, and some factors are still under discussion.

According to the analysis of single-layer and multiple layer (natural layer) oil test data of 4 wells in "someone" oil field in Sudan, the following three curves are obtained:

(1) In one group of development zones, when difference of thickness of all layers is less than 2 times, it can be seen from the trend of relationship curve between dimensionless oil production index and dimensionless effective thickness (Figure 1-22) that, before dimensionless effective thickness (cumulative effective thickness of multiple layer compared with effective thickness of the first layer) increases to 5 ($h_D < 5$), the dimensionless oil production index rises faster, and when $h_D > 5$, the dimensionless oil production index rises significantly slower. It shows that when the effective thickness in one group of development zones is not greater than 5 times of effective thickness of the first layer in this group, the interlayer interference is small. The expression is:

$$J_D = -0.0152h_D^2 + 0.2709h_D + 0.6455 \quad (1-6)$$

where J_D——dimensionless oil production index;

h_D——dimensionless effective thickness.

The prerequisite for application of this equation is $h_D < 9$.

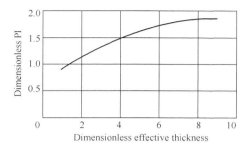

Figure 1−22 Relationship of dimensionless production index vs. dimensionless effective thickness

(2) From trend of the curve between number of layers in one group of development zones and dimensionless oil production index, it can be seen that when number of layers in one group of development zones is less than 5 ~ 7, the interlayer interference is smaller. This is more reasonable. As shown in Figure 1−23. The expression is:

$$J_\mathrm{D} = -0.0117i^2 + 0.2488i + 0.6682 \tag{1-7}$$

where J_D——dimensionless oil production index;

i——number of layers in the group of development zones.

The prerequisite for application of this equation is $i < 11$.

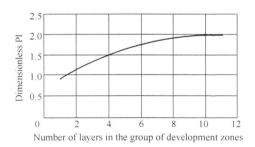

Figure 1−23 Relationship of dimensionless oil production index vs. number of layers

(3) For one group of development zones, the effective thickness should be less than 12m. If the physical properties of different layers in this group are very similar, the maximum effective thickness should be not more than 16m. So that interlayer interference within this group is small enough to assure this group of development zones be fully and effectively produced (Figure 1−24). The expression is:

$$J_\mathrm{D} = -0.0019h^2 + 0.0967h + 0.6455 \tag{1-8}$$

where, h is effective thickness of group of development zones, m. The prerequisite for application of this equation is $h < 26\text{m}$.

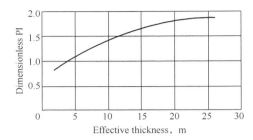

Figure 1-24 Relationship of dimensionless oil production index vs. effective thickness

The above three curves are calculated from the same data. The quantitative technical limits for dividing and grouping development zones are analyzed from different aspects. Among the above three parameters, the number of layers in one group is dominant, the other two are auxiliary parameters.

Besides the aforementioned three limits, single well controlled reserves, vertical interval of oil layers and difference in water saturation should also be considered when dividing and grouping development zones.

Section 5 Economic Evaluation

Economic evaluation is the work that based on the guidelines and principles of oilfield development, with the premise of ensuring the highest ultimate oil recovery, we select the oilfield development plan saving investment and having good economic benefit, so as to economize and accumulate funds.

1. Tasks of Economic Evaluation

The economic evaluation for oilfield development is to analyze the economic benefits of the oilfield development plans and provide basic information for decision-making on investment. In consistence with the characteristics of oil field development projects, the main tasks of economic evaluation on oilfield development include the following three aspects.

1.1 Carry out economic evaluation and feasibility study for engineering technical plan

The economic evaluation team should cooperate with departments of production management and designing to do a good job in the economic evaluation and feasibility study of the engineering technical plan, so as to provide the decision-making basis for improving the comprehensive economic benefits of engineering investment projects. The engineering technical plans mainly include: (1) oilfield development plan for a new area, (2) oilfield development adjustment plan for a historical area, (3) international cooperative oilfield development plan, and (4) economic evaluation on undeveloped reserves, etc.

1.2 Carry out marginal benefit analysis about oilfield development project

In order to analyze the economic limit of technical plans or technical measures, we carry out the marginal benefit analysis on oilfield development work. In practice, the problems to be analyzed are: (1) limit of oil productivity or limit of daily oil production rate per well, (2) reasonable well density, (3) limit of water cut of oil well, and (4) limit of steam-oil ratio for thermal recovery project.

1.3 Carry out predicting and analyzing the future economic performance of oilfield development plan

For a medium and long term oilfield development plan, the problems should be investigated are: (1) to predict the economic effect, and (2) to analyze the economics performance, etc.

2. Bases for Economic Evaluation

To carry out economic evaluation, we must calculate key economic indexes based on technical indexes and data of investment, cost, expense, etc.

2.1 Technical indexes of oilfield development performance

Technical indexes are parameters that characterize the dynamic response of oilfield production, including oil output, water cut, recovery percent, formation pressure, and cumulative production, etc.

Methods of calculating oilfield development indexes: laboratory experiment, reservoir numerical simulation, analytical method, empirical method, general mathematical method and logging and testing method.

The main technical indexes on which the economic analysis or calculation is based include (1) the arrangement of well pattern, especially the total number of wells drilled, the number of oil production wells and the number of water injection wells in the oilfield, (2) the oil production velocity, oil production rate and percentage of water cut increase during the oilfield development stage, (3) the number of years of every development stage and the total development period of years, (4) the number of wells using different production technologies in every development stage, namely, the number of flowing wells and the number of artificial lift wells, (5) water or gas injection plan, i.e., water or gas volume injected at every development stage, (6) recovery percent and estimated ultimate recovery factor at every development stage, and (7) main technological measures adopted in the development process, etc.

2.2 Parameters of investment, cost and expense

(1) Investment. The investments that needs to be predicted include exploration investment, drilling development wells, surface facilities construction, system engineering (crude oil storage and transportation, gas transportation, oil and gas processing, power supply, communication system, water supply and discharging, and roads, etc.), public engineering (machine repair and maintenance, logistics and auxiliary enterprises, civil construction in field area, other non-installation equipment, comprehensive utilization, environmental protection, and computers, etc.), interest during the construction period (calculated by compound interest rate to the end of construction period), and operating fund, etc.

(2) Oil and gas production cost. The oil and gas production cost are divided into 14 items, including material costs, workers' wages, employees' welfare, downhole operation costs, repair costs, power costs, other production costs, well logging and testing costs, fuel costs, oil and gas processing, water (gas) injection, light hydrocarbon collecting, depreciations and thermal driving of heavy oil.

(3) Fees.

①Management fees refer to the expenses of the administrative department of the enterprise for managing and organizing business activities, including company funds, labor union funds, employee education, labor insurance, unemployment insurance, directors' membership fees, consulting fees, audit fees, legal affairs, pollution exhaust, greening, taxes (property tax, vehicle and ship operating, stamp duty, land employment tax, etc.), land loss compensation fees, technology transfer fees, technology development fees, amortization of intangible assets, amortization of initial expenditure, amortization of business expenses, business entertainment expenses, bad debt losses, inventory losses, damage and abandonment (less inventory gains) and other management expenses.

②The mineral resources compensation fee should be calculated according to the national tolling regulations.

③Financial expenses refer to various expenses incurred by an enterprise to raise funds, including interest expenses during the period of production and operation of the enterprise (interest income being removed), net losses for currency exchange, charges for handling foreign exchange, charges by financial institutions, and other financial expenses incurred by funding activities.

④Sales expenses are estimated according to a certain proportion of income from sales.

3. Indexes for Economic Evaluation

Economic evaluation takes financial internal rate of return, investment payback period, financial net present value, etc., as the main appraising indexes. In addition, according to the characteristics and actual needs of the project, it may be necessary to calculate other indexes such as rate of return on investment, rate of profit and tax from investment, rate of profit on capital,

Chapter 1 Key Principles on Oilfield Development Engineering

repayment period of bank loan, ratio of current assets, ratio of quick assets, etc., in order to conduct auxiliary analyses.

(1) Financial internal rate of return. The financial internal rate of return refers to the discount rate at which the present value of the net cash flow of all years in the whole calculation period equals zero. It reflects the profit rate of the funds occupied by the project. It is the main dynamic evaluation index to investigate the profitability of the project, and its expression is

$$\sum_{t=1}^{n} (C_1 - C_0)_t (1 + FIRR)^{-t} = 0 \qquad (1-9)$$

where $FIRR$——financial internal rate of return;
   C_1——cash inflow;
   C_0——cash outflow;
   $(C_1 - C_0)_t$——net cash flow in year t;
   n——calculation period.

(2) Investment payback period. Investment payback period refers to the time required to offset all investments (fixed asset investment, adjustment tax for investment orientation, and operating capital) with the net income of the project. It is the main static appraising index to investigate the payback ability of financial investment on the project.

(3) Financial net present value. Financial net present value refers to the sum of the net cash flows of all years in the calculation period of the project, converted to the present values at the initial year of the calculation period, according to the primary standard of internal rate of return in petroleum industry or a specified discount rate. It is a dynamic index to examine the profitability of the project during the calculation period.

$$FNPV = \sum_{t=1}^{n} (C_1 - C_0)_t (1 + i_c)^{-t} \qquad (1-10)$$

where $FNPV$——financial net present value;
   i_c——annual discount rate.

Section 6 Oilfield Development Plan

The oilfield development plan is a programmatic document guiding the whole process of oilfield development, which scientifically demonstrates and clearly stipulates major issues such as grouping of development zones, development mechanisms, well pattern oil production velocity, etc.

A complete oilfield development plan includes at least four parts, i. e. , reservoir engineering, drilling engineering, oil production engineering and surface engineering.

The reservoir engineering plan mainly includes four aspects: oilfield overview, geology and reservoir characteristics, reservoir engineering design and requirements for implementation.

1. Oilfield Overview

The overview of the oilfield mainly presents the geographical location, climate, hydrology, transportation and economic conditions of the oilfield, describes the exploration history and degree of the oilfield, and introduces the preparation for oilfield development. Specifically, it includes the number and density of discovery wells and evaluation wells, seismic workload and processing technology, seismic line density and interpretation results, coring and analysis data, logging and explanation, formation test, appraisal trial production and development test, oilfield scope and oil-bearing layers, etc.

2. Geology and Reservoir Characteristics

Geology and reservoir characteristics include all kinds of features in structure, oil-bearing layers, reservoir beds, fluid, pressure and temperature system, porous flow physics, natural energy, and calculation and evaluation about reserves, etc.

3. Reservoir Engineering Design

Reservoir engineering design mainly includes the designs of group of development zones, development mechanism, development well pattern, oil production velocity, calculation of development index, and evaluation and optimization of the reservoir engineering plan, etc. Reservoir engineering design should adhere to the principle of "less investment, more revenue and maximum of economic benefit". Well pattern deployment should insist on the principle of "high yield with as smaller number of wells as possible". Oilfield development indexes are the predictive calculation results of oil production, water production, gas production and formation pressure, etc. , during a certain development period, of the design scheme. At present, reservoir numerical simulation method is generally used for calculation of technical indexes. Evaluation and optimization of the scheme is to calculate the economic benefit of all candidates based on the development indexes, according to petroleum industry standards, and then select the best scheme for implementation.

4. Requirements for Implementation of Reservoir Engineering Plan

According to the geological characteristics of the oilfield, the requirements for implementation of a reservoir engineering plan generally include: (1) sequence for drilling wells, methods for well completion, reservoir protection measures, sequence for putting wells into work, scheme for water injection and running timetable; (2) arrangement for pilot test of oilfield development and related requirements; (3) requirements for stimulation treatments; (4) dynamic monitoring requirements, including items and contents of dynamic monitoring; (5) requirements upon HSE and other aspects.

The detailed oilfield development plan also includes drilling engineering, production engineering and surface engineering, contents of which can refer to corresponding petroleum industry standards.

With a well designed oilfield development plan, there is a blueprint for oilfield development engineering, which can be implemented step by step according to the prescriptions in the plan.

Chapter 2 Method to Calculate Displacement Efficiency

Section 1 Definition of Recovery Factor

The recovery factor (efficiency) R_F of any secondary or tertiary oil recovery method is the product of a combination of three individual efficiency factors as given by the following generalized expression:

$$R_F = E_D E_A E_V \qquad (2-1)$$

$$N_p = N_S E_D E_A E_V \qquad (2-2)$$

where R_F——overall recovery factor;

E_D——displacement efficiency;

E_A——areal sweep efficiency;

E_V——vertical sweep efficiency;

N_S——initial oil in place at the start of the flood;

N_p——cumulative oil produced.

The major factors determining areal sweep are fluid mobilities, pattern type, areal heterogeneity and total volume of fluid injected.

The vertical sweep is primarily a function of vertical heterogeneity, degree of gravity segregation, fluid mobilities and total volume of injection.

As defined previously, displacement efficiency is the fraction of movable oil that has been recovered from the swept zone at any given time. Mathematically, the displacement efficiency is expressed as:

$$E_D = \frac{V_S - V_R}{V_S} \qquad (2-3)$$

Chapter 2 Method to Calculate Displacement Efficiency

$$E_D = \frac{V_p \dfrac{S_{oi}}{B_{oi}} - V_p \dfrac{\overline{S}_o}{B_o}}{V_p \dfrac{S_{oi}}{B_{oi}}} \tag{2-4}$$

where V_S——volume of oil at start of flood;

V_R——Remaining oil volume;

S_{oi}——initial oil saturation at start of flood;

B_{oi}——oil FVF at start of flood;

$\overline{S}_o$——average oil saturation in the flood pattern at a particular point during the flood.

Assuming a constant oil formation volume factor during the flood life, equation is reduced to:

$$E_D = \frac{S_{oi} - \overline{S}_o}{S_{oi}} \tag{2-5}$$

Where, the initial oil saturation S_{oi} is given by:

$$S_{oi} = 1 - S_{wi} - S_{gi} \tag{2-6}$$

However, in the swept area, the gas saturation is considered zero, thus:

$$\overline{S}_o = 1 - \overline{S}_w \tag{2-7}$$

The displacement efficiency E_D can be expressed more conveniently in terms of water saturation:

$$E_D = \frac{\overline{S}_w - S_{wi} - S_{gi}}{1 - S_{wi} - S_{gi}} \tag{2-8}$$

where $\overline{S}_w$——average water saturation in the swept area;

S_{gi}——initial gas saturation at the start of the flood;

S_{wi}——initial water saturation at the start of the flood.

If no initial gas is present at the start of the flood, equation is reduced to:

$$E_D = \frac{\overline{S}_w - S_{wi}}{1 - S_{wi}} \tag{2-9}$$

The displacement efficiency E_D will continually increase at different stages of the flood, i.e., with increasing $\overline{S}_w$, equation suggests that E_D reaches its maximum when the average oil saturation in the area of the flood pattern is reduced to the residual oil saturation S_{or} or, equivalently, when $\overline{S}_w = 1 - S_{or}$.

E_D will continually increase with increasing water saturation in the reservoir. The problem, of course, lies with developing an approach for determining the increase in the average water saturation in the swept area as a function of cumulative water injected (or injection time).

Buckley and Leverett (1942) developed a well-established theory, called the frontal displacement theory, which provides the basis for establishing such a relationship. This classic theory consists of two equations:

(1) Fractional flow equation.

(2) Frontal advance equation.

The frontal displacement theory and its main two components are discussed next.

Section 2 Fractional Flow Equation

1. The Development of the Fractional Flow Equation

The development of the fractional flow equation is attributed to Leverett (1941). For two immiscible fluids, oil and water, the fractional flow of water, f_w (or any immiscible driving fluid), is defined as the water flow rate divided by the total flow rate, or:

$$f_w = \frac{q_w}{q_t} = \frac{q_w}{q_w + q_o} \tag{2-10}$$

where f_w——fraction of water in the flowing stream, i.e., water cut;

q_t——total flow rate, bbl/d;

q_w——water flow rate, bbl/d;

q_o——oil flow rate, bbl/d.

Consider the steady-state flow of two immiscible fluids (oil and water) through a tilted-linear porous media (Figure 2-1). Assuming a homogeneous system, Darcy's equation can be applied for each of the fluids:

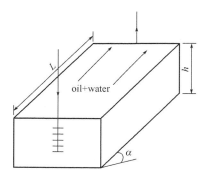

Figure 2-1 Linear displacement in a tilted system

$$q_o = \frac{-K_o A}{\mu_o} \left(\frac{\partial P_o}{\partial x} + g\rho_o \sin\alpha \right) \tag{2-11}$$

$$q_w = \frac{-K_w A}{\mu_w} \left(\frac{\partial P_w}{\partial x} + g\rho_w \sin\alpha \right) \tag{2-12}$$

where K_o, K_w ——effective permeability;

μ_o, μ_w ——viscosity;

P_o, P_w ——pressure;

ρ_o, ρ_w ——density;

A ——cross-sectional area;

x ——distance;

α ——dip angle, $\sin\alpha$ positive for updip flow and negative for downdip flow.

Subscripts o, w = oil and water.

$$f_w = \frac{1 + \dfrac{K_o A}{\mu_o q_t}\left(\dfrac{\partial P_c}{\partial x} - g\Delta\rho\sin\alpha\right)}{1 + \dfrac{K_o}{K_w}\dfrac{\mu_w}{\mu_o}} \quad (2-13)$$

where $\Delta\rho = \rho_w - \rho_o, P_c = P_o - P_w$

In field units, the above equation can be expressed as:

$$f_w = \frac{1 + \dfrac{0.001127 K_o A}{\mu_o q_t}\left(\dfrac{\partial P_c}{\partial x} - 0.433\Delta\rho\sin\alpha\right)}{1 + \dfrac{K_o}{K_w}\dfrac{\mu_w}{\mu_o}} \quad (2-14)$$

$$f_w = \frac{1 + \dfrac{0.001127 K K_{ro} A}{\mu_o i_w}\left(\dfrac{\partial P_c}{\partial x} - 0.433\Delta\rho\sin\alpha\right)}{1 + \dfrac{K_{ro}}{K_{rw}}\dfrac{\mu_w}{\mu_o}} \quad (2-15)$$

where i_w ——water injection rate, bbl/d;

K_{ro} ——relative permeability of oil;

K_{rw} ——relative permeability of water.

The fractional flow equation as expressed by the above relationship suggests that for a given rock fluid system, all the terms in the equation are defined by the characteristics of the reservoir, except water injection rate, i_w, water viscosity, μ_w and direction of the flow, i.e., updip or downdip injection.

Equation can be expressed in a more generalized form to describe the fractional flow of any displacing fluid as:

$$f_D = \frac{1 + \dfrac{0.001127 K K_{rD} A}{\mu_o i_D}\left(\dfrac{\partial P_c}{\partial x} - 0.433\Delta\rho\sin\alpha\right)}{1 + \dfrac{K_{ro}}{K_{rD}}\dfrac{\mu_D}{\mu_o}} \quad (2-16)$$

Where, the subscript D refers to the displacing fluid and $\Delta\rho$ is defined as:

$$\Delta\rho = \rho_D - \rho_o \quad (2-17)$$

For example, when the displacing fluid is immiscible gas, then:

$$f_g = \frac{1 + \dfrac{0.001127\, KK_{rg}A}{\mu_o i_g}\left[\dfrac{\partial P_c}{\partial x} - 0.433(\rho_g - \rho_o)\sin\alpha\right]}{1 + \dfrac{K_{ro}}{K_{rg}}\dfrac{\mu_g}{\mu_o}} \quad (2-18)$$

The effect of capillary pressure is usually neglected because the capillary pressure gradient is generally small:

$$f_w = \frac{1 + \dfrac{0.001127\, KK_{ro}A}{\mu_o i_w}\Big)\Big(0.433\Delta\rho\sin\alpha\Big)}{1 + \dfrac{K_{ro}}{K_{rw}}\dfrac{\mu_w}{\mu_o}} \quad (2-19)$$

$$f_g = \frac{1 + \dfrac{0.001127\, KK_{rg}A}{\mu_o i_g}\left[0.433(\rho_g - \rho_o)\sin\alpha\right]}{1 + \dfrac{K_{ro}}{K_{rg}}\dfrac{\mu_g}{\mu_o}} \quad (2-20)$$

From the definition of water cut, i.e., $f_w = q_w/(q_w + q_o)$, we can see that the limits of the water cut are 0 and 100%. At the irreducible (connate) water saturation, the water flow rate q_w is zero and, therefore, the water cut is 0%. At the residual oil saturation point, S_{or}, the oil flow rate is zero and the water cut reaches its upper limit of 100%. The shape of the water cut versus water saturation curve is characteristically S-shaped, as shown in Figure 2-2. The limits of the f_w curve (0 and 1) are defined by the end points of the relative permeability curves.

2. The Effect Factors of the Fractional Flow Equation

2.1 Effect of water and oil viscosities

Figure 2-3 shows the general effect of oil viscosity on the fractional flow curve for both water-wet and oil-wet rock systems. This illustration reveals that regardless of the system wettability, a higher oil viscosity results in an upward shift (an increase) in the fractional flow curve.

The apparent effect of the water viscosity on the water fractional flow is clearly indicated by examining equation. Higher injected water viscosities will result in an increase in the value of the denominator of equation with an overall reduction in f_w (i.e., a downward shift).

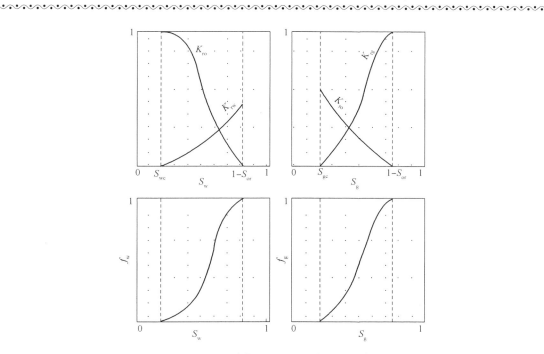

Figure 2 – 2 Fractional flow curves as function of saturations

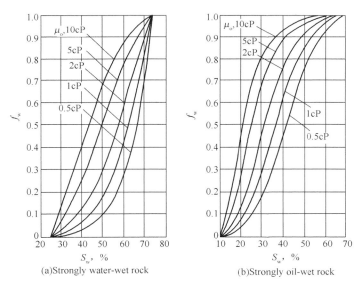

Figure 2 – 3 Effect of oil viscosity on f_w

2.2 Effect of dip angle and injection rate

To study the effect of the formation dip angle α and the injection rate on the displacement efficiency, consider the water fractional flow equation as represented by equation. Assuming a constant injection rate and realizing that $(\rho_w - \rho_o)$ is always positive and in order to isolate the effect of the dip angle and injection rate on f_w, equation is expressed in the following simplified form.

$$f_w = \frac{1 - X\dfrac{\sin\alpha}{i_w}}{1 + Y} \qquad (2-21)$$

Where, the variables X and Y are a collection of different terms that are all considered positives and given by:

$$X = \frac{0.001127 \times 0.433 \times KK_{ro}A(\rho_w - \rho_o)}{\mu_o} \qquad (2-22)$$

$$Y = \frac{K_{ro}}{K_{rw}}\frac{\mu_w}{\mu_o} \qquad (2-23)$$

Updip flow, i.e. $\sin\alpha$ is positive. Figure 2-4 shows that when the water displaces oil updip (i.e., injection well is located downdip), a more efficient performance is obtained. This improvement is due to the fact that the term $X\sin\alpha/i_w$ will always remain positive, which leads to a decrease (downward shift) in the f_w curve. Equation (2-21) reveals that a lower water-injection rate i_w is desirable since the numerator $1-X\sin\alpha/i_w$ of equation will decrease with a lower injection rate i_w, resulting in an overall downward shift in the f_w curve.

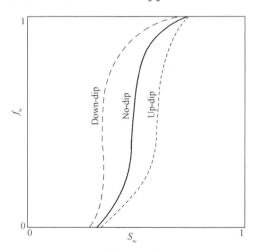

Figure 2-4 Effect of dip angle on f_w

Downdip flow, i.e., $\sin\alpha$ is negative. When the oil is displaced downdip (i.e., injection well is located updip), the term $X\sin\alpha/i_w$ will always remain negative and, therefore, the numerator of equation will be $1 + X\sin\alpha/i_w$, i.e.:

$$f_w = \frac{1 + X\dfrac{\sin\alpha}{i_w}}{1 + Y} \qquad (2-24)$$

which causes an increase (upward shift) in the f_w curve. It is beneficial, therefore, when injection wells are located at the top of the structure to inject the water at a higher injection rate to improve the displacement efficiency.

It is interesting to reexamine equation, when displacing the oil downdip. Combining the product $X \sin\alpha$ as C, equation can be written as:

$$f_w = \frac{1 + \dfrac{C}{i_w}}{1 + Y} \qquad (2-25)$$

The above expression shows that the possibility exists that the water cut f_w could reach a value greater than unity ($f_w > 1$) if:

$$\frac{C}{i_w} > Y \qquad (2-26)$$

This could only occur when displacing the oil downdip at a low water-injection rate i_w. The resulting effect of this possibility is called a counter flow, where the oil phase is moving in a direction opposite to that of the water (i.e., oil is moving upward and the water downward). When the water injection wells are located at the top of a tilted formation, the injection rate must be high to avoid oil migration to the top of the formation.

Note that for a horizontal reservoir, i.e., $\sin\alpha = 0$, the injection rate has no effect on the fractional flow curve. When the dip angle α is zero, equation is reduced to the following simplified form:

$$f_w = \frac{1}{1 + \dfrac{K_{ro}}{K_{rw}} \dfrac{\mu_w}{\mu_o}} \qquad (2-27)$$

3. Relationship Between Water Cut and Water-oil Ratio

3.1 Reservoir f_w reservoir WOR_r relationship

$$f_w = \frac{q_w}{q_w + q_o} = \frac{\left(\dfrac{q_w}{q_o}\right)}{\left(\dfrac{q_w}{q_o}\right) + 1} \qquad (2-28)$$

Substituting for WOR gives:

$$f_w = \frac{WOR_r}{WOR_r + 1} \qquad (2-29)$$

Solving for WOR_r gives:

$$WOR_r = \frac{1}{\frac{1}{f_w} - 1} = \frac{f_w}{1 - f_w} \qquad (2-30)$$

3.2 Reservoir f_w surface WOR_s relationship

By definition:

$$f_w = \frac{q_w}{q_w + q_o} = \frac{Q_w B_w}{Q_w B_w + Q_o B_o} = \frac{\frac{Q_w}{Q_o} B_w}{\frac{Q_w}{Q_o} B_w + B_o} \qquad (2-31)$$

Introducing the surface WOR_s into the above expression gives:

$$f_w = \frac{B_w WOR_s}{B_w WOR_s + B_o} \qquad (2-32)$$

Solving for WOR_s yields:

$$WOR_s = \frac{B_o}{B_w \left(\frac{1}{f_w} - 1\right)} = \frac{B_o f_w}{B_w (1 - f_w)} \qquad (2-33)$$

3.3 Reservoir WOR_r surface WOR_s relationship

From the definition of WOR:

$$WOR_r = \frac{q_w}{q_o} = \frac{Q_w B_w}{Q_o B_o} = \frac{\frac{Q_w}{Q_o} B_w}{B_o} \qquad (2-34)$$

Introducing the surface WOR_s into the above expression gives:

$$WOR_r = WOR_s \frac{B_w}{B_o} \qquad (2-35)$$

or

$$WOR_s = WOR_r \frac{B_o}{B_w} \qquad (2-36)$$

3.4 Surface f_{ws} surface WOR_s relationship

$$f_{ws} = \frac{Q_w}{Q_w + Q_o} = \frac{\frac{Q_w}{Q_o}}{\frac{Q_w}{Q_o} + 1} \qquad (2-37)$$

$$f_{ws} = \frac{WOR_s}{WOR_s} + 1 \qquad (2-38)$$

3.5 Surface f_{ws} reservoir f_w relationship

$$f_{ws} = \frac{B_o}{B_w \left(\dfrac{1}{f_w} - 1\right) + B_o} \qquad (2-39)$$

The fractional flow equation, as discussed in the previous section, is used to determine the water cut f_w at any point in the reservoir, assuming that the water saturation at the point is known.

The question, however, is how to determine the water saturation at this particular point. The answer is to use the frontal advance equation. The frontal advance equation is designed to determine the water saturation profile in the reservoir at any given time during water injection.

Section 3   Frontal Advance Equation

Buckley and Leverett (1942) presented what is recognized as the basic equation for describing two-phase, immiscible displacement in a linear system.

The equation is derived based on developing a material balance for the displacing fluid as it flows through any given element in the porous media:

Volume entering the element-Volume leaving the element = change in fluid volume

$$(2-40)$$

Consider a differential element of porous media, as shown in Figure 2 – 5, having a differential length dx, an area A, and a porosity ϕ.

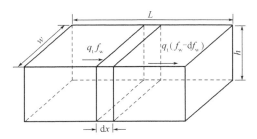

Figure 2 – 5 Water flow through a linear differential element

During a differential time period dt, the total volume of water entering the element is given by:

Volume of water entering the element = $q_t f_w \mathrm{d}t$

The volume of water leaving the element has a differentially smaller water cut $(f_w - \mathrm{d}f_w)$ and is given by:

Volume of water leaving the element $= q_t(f_w - df_w)dt$

Subtracting the above two expressions gives the accumulation of the water volume within the element in terms of the differential changes of the saturation df_w:

$$q_t f_w dt - q_t(f_w - df_w)dt = A\phi dx\, dS_w/5.615 \qquad (2-41)$$

Simplifying:

$$q_t df_w dt = A\phi\, dx\, dS_w/5.615 \qquad (2-42)$$

Separating the variables gives:

$$\left(\frac{dx}{dt}\right)_{S_w} = v_{S_w} = \frac{5.615 q_t}{\phi A}\left(\frac{df_w}{dS_w}\right)_{S_w} \qquad (2-43)$$

where $(v)_{S_w}$——velocity of any specified value of S_w, ft/day;

A——sectional area, ft^2;

q_t——total flow rate (oil + water), bbl/d;

$(df_w/dS_w)_{S_w}$——slope of the fw vs. S_w curve at S_w.

The above relationship suggests that the velocity of any specific water saturation S_w is directly proportional to the value of the slope of the f_w vs. S_w curve, evaluated at S_w. Note that for two-phase flow, the total flow rate q_t is essentially equal to the injection rate i_w, or:

$$\left(\frac{dx}{dt}\right)_{S_w} = (v)_{S_w} = \frac{5.615 i_w}{\phi A}\left(\frac{df_w}{ds_w}\right)_{S_w} \qquad (2-44)$$

To calculate the total distance any specified water saturation will travel during a total time t, equation must be integrated,

$$\int_0^x dx = \frac{5.615 i_w}{\phi A}\left(\frac{df_w}{dS_w}\right)_{S_w}\int_0^t dt \qquad (2-45)$$

or

$$(x)_{S_w} = \frac{5.615 t i_w}{\phi A}\left(\frac{df_w}{dS_w}\right)_{S_w} \qquad (2-46)$$

Equation can also be expressed in terms of total volume of water injected by recognizing that under a constant water injection rate, the cumulative water injected is given by:

$$W_{inj} = t i_w \qquad (2-47)$$

or

$$(x)_{S_w} = \frac{5.615 W_{inj}}{\phi A}\left(\frac{df_w}{dS_w}\right)_{S_w} \qquad (2-48)$$

where　W_{inj}——cumulative injected water, bbl;

　　t——time, d;

　　$(x)_{S_w}$——distance from the injection for any given saturation S_w, ft.

Equation also suggests that the position of any value of water saturation S_w at given cumulative water injected W_{inj} is proportional to the slope (df_w/dS_w) for this particular S_w. At any given time t, the water saturation profile can be plotted by simply determining the slope of the f_w curve at each selected saturation and calculating the position of S_w from equation.

Figure 2–6 shows the typical S shape of the f_w curve and its derivative curve. However, a mathematical difficulty arises when using the derivative curve to construct the water saturations profile at any given time. Suppose we want to calculate the positions of two different saturations (shown in Figure as saturations A and B) after W_{inj} barrels of water have been injected in the reservoir. Applying equation gives:

$$(x)_A = \frac{5.615 W_{inj}}{\phi A} \left(\frac{df_w}{dS_w}\right)_A \quad (2-49)$$

$$(x)_B = \frac{5.615 W_{inj}}{\phi A} \left(\frac{df_w}{dS_w}\right)_B \quad (2-50)$$

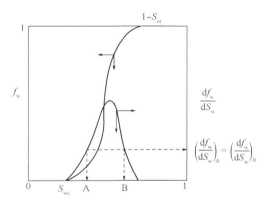

Figure 2–6　The f_w curve with its saturation derivative curve

Figure 2–6 indicates that both derivatives are identical, i. e., $(df_w/dS_w)_A = (df_w/dS_w)_B$, which implies that multiple water saturations can coexist at the same position, but this is physically impossible. Buckley and Leverett (1942) recognized the physical impossibility of such a condition. They pointed out that this apparent problem is due to the neglect of the capillary pressure gradient term in the fractional flow equation. This capillary term is given by:

$$\text{Capillary term} = \frac{0.001127 K_o A}{\mu_o i_w} \frac{dP_c}{dx}$$

Including the above capillary term when constructing the fractional flow curve would produce a graphical relationship that is characterized by the following two segments of lines, as shown in Figure 2-7: A straight line segment with a constraint slope of $(df_w/dS_w)_{S_{wf}}$ from S_{wc} to S_{wf}; A concaving curve with decreasing slopes from S_{wf} to $(1-S_{or})$.

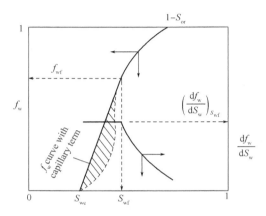

Figure 2-7   Effect of the capillary term on the f_w curve

Terwilliger et al. (1951) found that at the lower range of water saturations between S_{wc} and S_{wf}, all saturations move at the same velocity as a function of time and distance. Notice that all saturations in that range have the same value for the slope and, therefore, the same velocity as given by equation.

$$(v)_{S_w < S_{wf}} = \frac{5.615 i_w}{\phi A} \left(\frac{df_w}{dS_w}\right)_{S_{wf}} \quad (2-51)$$

We can also conclude that all saturations in this particular range will travel the same distance x at any particular time, as given by Equation.

$$(x)_{S_w < S_{wf}} = \frac{5.615 i_w t}{\phi A} \left(\frac{df_w}{dS_w}\right)_{S_{wf}} \quad (2-52)$$

The result is that the water saturation profile will maintain a constant shape over the range of saturations between S_{wc} and S_{wf} with time. Terwilliger and his coauthors termed the reservoir-flooded zone with this range of saturations the stabilized zone. They define the stabilized zone as that particular saturation interval (i.e., S_{wc} to S_{wf}) where all points of saturation travel at the same velocity. Figure 2-8 illustrates the concept of the stabilized zone. The authors also identified another saturation zone between S_{wf} and $(1-S_{or})$, where the velocity of any water saturation is variable. They termed this zone the nonstabilized zone.

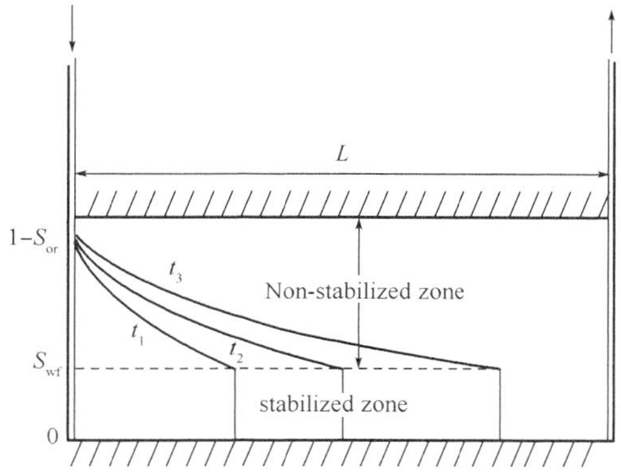

Chapter 2 Method to Calculate Displacement Efficiency

Figure 2 – 8 Water saturation profile as a function of distance and time

Experimental core flood data show that the actual water saturation profile during water flooding is similar to that of Figure 2 – 8. There is a distinct front, or shock front, at which the water saturation abruptly increases from S_{wc} to S_{wf}. Behind the flood front there is a gradual increase in saturations from S_{wf} up to the maximum value of $1 - S_{or}$. Therefore, the saturation S_{wf} is called the water saturation at the front or, alternatively, the water saturation of the stabilized zone.

Welge (1952) showed that by drawing a straight line from S_{wc} (or from S_{wi} if it is different from S_{wc}) tangent to the fractional flow curve, the saturation value at the tangent point is equivalent to that at the front S_{wf}. The coordinate of the point of tangency represents also the value of the water cut at the leading edge of the water front f_{wf}.

The water saturation profile at any given time t_1 can be easily developed as follows:

Step 1. Ignoring the capillary pressure term, construct the fractional flow curve, i. e., f_w vs. S_w.

Step 2. Draw a straight line tangent from S_{wi} to the curve.

Step 3. Identify the point of tangency and read off the values of S_{wf} and f_{wf}.

Step 4. Calculate graphically the slope of the tangent as $(df_w/dS_w)_{S_{wf}}$.

Step 5. Calculate the distance of the leading edge of the water front from the injection well by using equation:

$$(x)_{S_{wf}} = \frac{5.615 i_w t_1}{\phi A} \left(\frac{df_w}{dS_w}\right)_{S_{wf}} \tag{2-53}$$

Step 6. Select several values for water saturation S_w greater than S_{wf} and determine $(df_w/dS_w)_{S_w}$ by graphically drawing a tangent to the f_w curve at each selected water saturation (as shown in Figure 2 – 9).

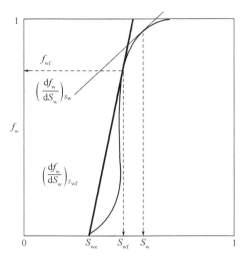

Figure 2-9 Fractional flow curve

Step 7. Calculate the distance from the injection well to each selected saturation by applying equation, or:

$$(x)_{S_w} = \frac{5.615 i_w t_1}{\phi A} \left(\frac{df_w}{dS_w}\right)_{S_w} \tag{2-54}$$

Step 8. Establish the water saturation profile after t_1 days by plotting results obtained in step 7.

Step 9. Select a new time t_2 and repeat steps 5 through 7 to generate a family of water saturation profiles as shown schematically in figure.

Some erratic values of $(df_w/dS_w)_{S_w}$ might result when determining the slope graphically at different saturations. A better way is to determine the derivative mathematically by recognizing that the relative permeability ratio (K_{ro}/K_{rw}) can be expressed as:

$$\frac{K_{ro}}{K_{rw}} = a e^{bS_w} \tag{2-55}$$

Notice that the slope b in the above expression has a negative value. The above expression can be substituted into fractional flow equation to give:

$$f_w = \frac{1}{1 + \frac{\mu_w}{\mu_o} a e^{bS_w}} \tag{2-56}$$

The derivative of $(df_w/dS_w)_{S_w}$ may be obtained mathematically by differentiating the above equation with respect to S_w to give:

$$\left(\frac{df_w}{dS_w}\right)_{S_w} = \frac{-\frac{\mu_w}{\mu_o} a b e^{bS_w}}{\left(1 + \frac{\mu_w}{\mu_o} a e^{bS_w}\right)^2} \tag{2-57}$$

Chapter 2 Method to Calculate Displacement Efficiency

The water front (leading edge) will eventually reach the production well and water breakthrough occurs.

At water breakthrough, the leading edge of the water front would have traveled exactly the entire distance between the two wells. Therefore, to determine the time to breakthrough, t_{BT}, simply set $(x)_{S_{wf}}$ equal to the distance between the injector and producer L and solve for the time:

$$L = \frac{5.615 i_w t_{BT}}{\phi A} \left(\frac{df_w}{dS_w}\right)_{S_{wf}} \tag{2-58}$$

Note that, the pore volume (PV) is given by:

$$PV = \frac{\phi AL}{5.615} \tag{2-59}$$

Combining the above two expressions and solving for the time to breakthrough t_{BT} gives:

$$t_{BT} = \frac{PV}{i_w} \frac{1}{\left(\frac{df_w}{dS_w}\right)_{S_{wf}}} \tag{2-60}$$

Assuming a constant water-injection rate, the cumulative water injected at breakthrough is calculated as:

$$W_{iBT} = i_w t_{BT} = \frac{PV}{\left(\frac{df_w}{dS_w}\right)_{S_{wf}}} \tag{2-61}$$

where W_{iBT}——cumulative water injected at breakthrough, bbl.

It is convenient to express the cumulative water injected in terms of pore volumes injected, i.e., dividing W_{inj} by the reservoir total pore volume.

Conventionally, Q_i refers to the total pore volumes of water injected. Q_i at breakthrough is:

$$Q_{iBT} = \frac{W_{iBT}}{PV} = \frac{1}{\left(\frac{df_w}{dS_w}\right)_{S_{wf}}} \tag{2-62}$$

where Q_{iBT}——cumulative pore volumes of water injected at breakthrough.

A further discussion is needed to better understand the significance of the Buckley and Leverett (1942) frontal advance theory. Equation which represents cumulative water injected at breakthrough is given by:

$$W_{iBT} = PV \frac{1}{\left(\frac{df_w}{dS_w}\right)_{S_{wf}}} = PV \, Q_{iBT} \tag{2-63}$$

If the tangent to the fractional flow curve is extrapolated to $f_w = 1$ with a corresponding water saturation of S_w^* (as shown in Figure 2-10), then the slope of the tangent can be calculated numerically as:

$$\left(\frac{df_w}{dS_w}\right)_{S_{wf}} = \frac{1-0}{S_w^* - S_{wi}} \qquad (2-64)$$

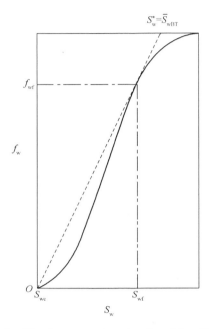

Figure 2-10 Average water saturation at breakthrough

Combining the above two expressions gives:

$$W_{iBT} = PV(S_w^* - S_{wi}) = PVQ_{iBT} \qquad (2-65)$$

The above equation suggests that the water saturation value denoted as S_w^* must be the average water saturation at breakthrough, or:

$$W_{iBT} = PV(\overline{S}_{wBT} - S_{wi}) = PVQ_{iBT} \qquad (2-66)$$

where $\overline{S}_{wBT}$——average water saturation in the reservoir at breakthrough.

Two important points must be considered when determining $\overline{S}_{wBT}$:

(1) When drawing the tangent, the line must be originated from the initial water saturation S_{wi} if it is different from the connate water saturation S_{wc}, as shown in Figure 2-11.

(2) When considering the areal sweep efficiency E_A and vertical sweep efficiency E_V, equation should be expressed as:

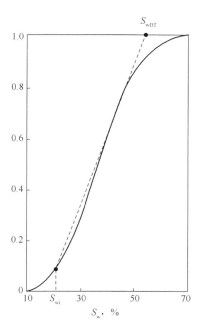

Figure 2 – 11 Tangent from S_{wi}

$$W_{iBT} = PV(\overline{S}_{wBT} - S_{wi})E_{ABT}E_{VBT} \qquad (2-67)$$

or equivalently as:

$$W_{iBT} = PVQ_{iBT}E_{ABT}E_{VBT} \qquad (2-68)$$

where E_{ABT}, E_{VBT}——the areal and vertical sweep efficiencies at breakthrough.

Note that the average water saturation in the swept area would remain constant with a value of $\overline{S}_{wBT}$ until breakthrough occurs, as illustrated in Figure 2 – 12. At the time of breakthrough, the flood front saturation S_{wf} reaches the producing well and the water cut increases suddenly from zero to f_{wf}. At break-through, S_{wf} and f_{wf} are designated $\overline{S}_{wBT}$ and f_{wBT}.

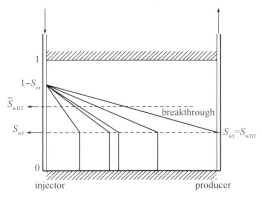

Figure 2 – 12 Average water saturation before breakthrough

After breakthrough, the water saturation and the water cut at the producing well gradually increase with continuous injection of water, as shown in Figure 2−13. Traditionally, the produced well is designated as well 2 and, therefore, the water saturation and water cut at the producing well are denoted as S_{w2} and f_{w2}, respectively.

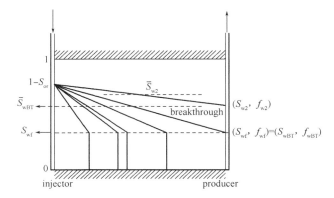

Figure 2−13 Average water saturation after breakthrough

Welge (1952) illustrated that when the water saturation at the producing well reaches any assumed value S_{w2} after breakthrough, the fractional flow curve can be used to determine:

(1) Producing water cut f_{w2};

(2) Average water saturation in the reservoir $\bar{S}_{w2}$;

(3) Cumulative water injected in pore volumes, i.e., Q_i.

As shown in Figure 2−14, the author pointed out that drawing a tangent to the fractional flow curve at any assumed value of S_{w2} greater than S_{wf} has the following properties:

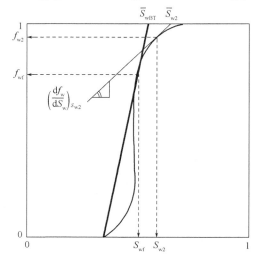

Figure 2−14 Determination of average water saturation after breakthrough

Chapter 2 Method to Calculate Displacement Efficiency

(1) The value of the fractional flow at the point of tangency corresponds to the well producing water cut f_{w2}, as expressed in bbl/bbl.

(2) The saturation at which the tangent intersects $f_w = 1$ is the average water saturation $\bar{S}_{w2}$ in the swept area. Mathematically, the average water saturation is determined from:

$$\bar{S}_{w2} = S_{w2} + \frac{1 - f_{w2}}{\left(\dfrac{df_w}{dS_w}\right)_{S_{w2}}} \tag{2-69}$$

(3) The reciprocal of the slope of the tangent is defined as the cumulative pore volumes of water injected Q_i at the time when the water saturation reaches S_{w2} at the producing well.

$$Q_i = \frac{1}{\left(\dfrac{df_w}{dS_w}\right)_{S_{w2}}} \tag{2-70}$$

(4) The cumulative water injected when the water saturation at the producing well reaches S_{w2} is given by:

$$W_{inj} = PV Q_i E_A E_V \tag{2-71}$$

or equivalently as:

$$W_{inj} = PV(\bar{S}_{w2} - S_{wi}) E_A E_V \tag{2-72}$$

For a constant injection rate i_w, the total time t to inject W_{inj} barrels of water is given by:

$$t = \frac{W_{inj}}{i_w} \tag{2-73}$$

Section 4 Oil Recovery Calculations

The main objective of performing oil recovery calculations is to generate a set of performance curves under a specific water-injection scenario.

A set of performance curves is defined as the graphical presentation of the time-related oil recovery calculations in terms of:

(1) Oil production rate, Q_o;

(2) Water production rate, Q_w;

(3) Surface water-oil ratio, WOR_s;

(4) Cumulative oil production, N_p;

(5) Recovery factor, R_F;

(6) Cumulative water production, W_p;

(7) Cumulative water injected, W_{inj};

(8) Water-injection pressure, p_{inj};

(9) Water-injection rate, i_w.

1. The Index of Oil Recovery Calculation

In general, oil recovery calculations are divided into two parts: (1) before breakthroughcalculations and (2) after breakthrough calculations. Regardless of the stage of the waterflood, i.e., before or after breakthrough, the cumulative oil production is given previously by equation as:

$$N_p = N_S E_D E_A E_V \qquad (2-74)$$

As defined by equation when $S_{gi} = 0$, the displacement efficiency is given by:

$$E_D = \frac{\bar{S}_w - S_{wi}}{1 - S_{wi}} \qquad (2-75)$$

At breakthrough, the E_D can be calculated by determining the average water saturation at breakthrough:

$$E_{DBT} = \frac{\bar{S}_{wBT} - S_{wi}}{1 - S_{wi}} \qquad (2-76)$$

where E_{DBT}——displacement efficiency at breakthrough;

S_{wBT}——average water saturation at breakthrough.

The cumulative oil production at breakthrough is then given by:

$$(N_p)_{BT} = N_S E_{DBT} E_{ABT} E_{VBT} \qquad (2-77)$$

where $(N_p)_{BT}$——cumulative oil production at breakthrough.

Assuming E_A and E_V are 100%, Equation is reduced to:

$$(N_p)_{BT} = N_S E_{DBT} \qquad (2-78)$$

2. The Step of Oil Recovery Calculation

2.1 Before breakthrough

Before breakthrough occurs, the oil recovery calculations are simple when assuming that no free gas exists at the start of the flood, i.e., $S_{gi} = 0$. The cumulative oil production is simply equal to the volume of water injected with no water production during this phase ($W_p = 0$ and $Q_w = 0$).

2.2 After breakthrough

Oil recovery calculations after breakthrough are based on determining E_D at various assumed

Chapter 2 Method to Calculate Displacement Efficiency

values of water saturations at the producing well. The specific steps of performing complete oil recovery calculations are composed of three stages: Data preparation, Recovery performance to breakthrough and Recovery performance after breakthrough.

Stage 1: Data Preparation

Step 1. Express the relative permeability data as relative permeability ratio K_{ro}/K_{rw} and plot their values versus their corresponding water saturations on a semi-log scale.

Step 2. Assuming that the resulting plot of relative permeability ratio, K_{ro}/K_{rw} vs. S_w, forms a straight-line relationship, determine values of the coefficients a and b of the straight line.

Express the straight-line relationship in the form given by equation, or:

$$\frac{K_{ro}}{K_{rw}} = a e^{bS_w} \qquad (2-79)$$

Step 3. Calculate and plot the fractional flow curve f_w, allowing for gravity effects if necessary, but neglecting the capillary pressure gradient.

Step 4. Select several values of water saturations between S_{wf} and $(1-S_{or})$ and determine the slope (df_w/dS_w) at each saturation. The numerical calculation of each slope as expressed by Equation provides consistent values as a function of saturation, or:

$$\left(\frac{df_w}{dS_w}\right)_{S_w} = \frac{-\frac{\mu_w}{\mu_o} a e^{bS_w}}{\left(1 + \frac{\mu_w}{\mu_o} a e^{bS_w}\right)^2} \qquad (2-80)$$

Step 5. Prepare a plot of the calculated values of the slope (df_w/dS_w) versus S_w on a Cartesian scale and draw a smooth curve through the points.

Stage 2: Recovery Performance to Breakthrough ($S_{gi}=0, E_A, E_V=100\%$)

Step 1. Draw a tangent to the fractional flow curve as originated from S_{wi} and determine:

(1) Point of tangency with the coordinate (S_{wf}, f_{wf});

(2) Average water saturation at breakthrough the tangent line to $f_w=1.0$;

(3) Slope of the tangent line $(df_w/dS_w)_{S_{wf}}$.

Step 2. Calculate pore volumes of water injected at breakthrough by using Equation:

$$Q_{iBT} = \frac{1}{\left(\dfrac{df_w}{dS_w}\right)_{S_{wf}}} = \overline{S}_{wBT} - S_{wi} \qquad (2-81)$$

Step 3. Assuming E_A and E_V are 100%, calculate cumulative water injected at breakthrough by applying Equation:

$$W_{iBT} = PV(\overline{S}_{wBT} - S_{wi}) \qquad (2-82)$$

or equivalently:
$$W_{iBT} = PVQ_{iBT} \qquad (2-83)$$

Step 4. Calculate the displacement efficiency at breakthrough by applying Equation:
$$E_{DBT} = \frac{\overline{S}_{wBT} - S_{wi}}{1 - S_{wi}} \qquad (2-84)$$

Step 5. Calculate cumulative oil production at breakthrough from Equation:
$$(N_p)_{BT} = N_S E_{DBT} \qquad (2-85)$$

Step 6. Assuming a constant water-injection rate, calculate time to breakthrough from Equation:
$$t_{BT} = \frac{W_{iBT}}{i_w} \qquad (2-86)$$

Step 7. Select several values of injection time less than the breakthrough time, i.e., $t < t_{BT}$, and set:
$$W_{inj} = i_w t$$
$$Q_o = i_w / B_o$$
$$WOR = 0$$
$$W_p = 0$$
$$N_p = \frac{i_w t}{B_o} = \frac{W_{inj}}{B_o} \qquad (2-87)$$

Step 8. Calculate the surface water-oil ratio WOR_s exactly at breakthrough by using Equation:
$$WOR_s = \frac{B_o}{B_w \left(\dfrac{1}{f_{wBT}} - 1 \right)} \qquad (2-88)$$

Note that WOR_s as calculated from the above expression is only correct when both the areal sweep efficiency E_A and vertical sweep efficiency E_V are 100%.

Stage 3: Recovery Performance after Breakthrough ($S_{gi}=0, E_A, E_V = 100\%$)

The recommended methodology of calculating recovery performance after breakthrough is based on selecting several values of water saturations around the producing well, i.e., S_{w2}, and determining the corresponding average reservoir water saturation $\overline{S}_{w2}$ for each S_{w2}. The specific steps that are involved are summarized below.

Step 1. Select six to eight different values of S_{w2} (i.e., S_w at the producing well) between S_{wBT} and $(1 - S_{or})$ and determine (df_w/dS_w) values corresponding to these S_{w2} points.

Step 2. For each selected value of S_{w2}, calculate the corresponding reservoir water cut and

Chapter 2 Method to Calculate Displacement Efficiency

average water saturation from Equations:

$$f_{w2} = \frac{1}{1 + \frac{\mu_w}{\mu_o} a e^{bS_{w2}}} \tag{2-89}$$

$$\overline{S}_{w2} = S_{w2} + \frac{1 - f_{w2}}{\left(\frac{df_w}{dS_w}\right)_{S_{w2}}} \tag{2-90}$$

Step 3. Calculate the displacement efficiency E_D for each selected value of S_{w2}:

$$E_D = \frac{\overline{S}_{w2} - S_{wi}}{1 - S_{wi}} \tag{2-91}$$

Step 4. Calculate cumulative oil production N_p for each selected value of S_{w2} from Equation, or:

$$N_p = N_S E_D E_A E_V \tag{2-92}$$

Assuming E_A and E_V are equal to 100%, then:

$$N_p = N_S E_D \tag{2-93}$$

Step 5. Determine pore volumes of water injected, Q_i, for each selected value of S_{w2} from Equation:

$$Q_i = \frac{1}{\left(\frac{df_w}{dS_w}\right)_{S_{w2}}} \tag{2-94}$$

Step 6. Calculate cumulative water injected for each selected value of S_{w2} by applying Equation:

$$W_{inj} = PVQ_i \text{ or } W_{inj} = PV(\overline{S}_{w2} - S_{wi}) \tag{2-95}$$

Notice that E_A and E_V are set equal to 100%

Step 7. Assuming a constant water-injection rate i_w, calculate the time t to inject W_{inj} barrels of water by applying Equation:

$$t = \frac{W_{inj}}{i_w} \tag{2-96}$$

Step 8. Calculate cumulative water production W_p at any time t from the material balance equation, which states that the cumulative water injected at any time will displace an equivalent volume of oil and water, or:

$$W_{inj} = N_p B_o + W_p B_w \tag{2-97}$$

Solving for W_p gives:

$$W_p = \frac{W_{inj} - N_p B_o}{B_w} \tag{2-98}$$

or equivalently in a more generalized form:

$$W_p = \frac{W_{inj} - (\bar{S}_{w2} - S_{wi})PVE_A E_V}{B_w} \quad (2-99)$$

We should emphasize that all of the above derivations are based on the assumption that no free gas exists from the start of the flood till abandonment.

Step 9. Calculate the surface water-oil ratio WOR_s that corresponds to each value of f_{w2} (as determined in step 2) from Equation:

$$WOR_s = \frac{B_o}{B_w\left(\dfrac{1}{f_{w2}} - 1\right)} \quad (2-100)$$

Step 10. Calculate the oil and water flow rates from the following derived relationships:

$$i_w = Q_o B_o + Q_w B_w \quad (2-101)$$

Introducing the surface water-oil ratio into the above expression gives:

$$i_w = Q_o B_o + Q_o B_w WOR_s \quad (2-102)$$

Solving for Q_o gives:

$$Q_o = \frac{i_w}{B_o + B_w WOR_s} \quad (2-103)$$

and

$$Q_w = Q_o WOR_s \quad (2-104)$$

Step 11. The preceding calculations as described in steps 1 through 10 can be organized in Table 2-1.

Table 2-1 The preceding calculations

S_{W2}	f_{W2}	df_w/dS_w	$\bar{S}_{w2}$	E_D	N_p	Q_i	W_{inj}	t	W_p	WOR_s	Q_o	Q_w
S_{wBT}	f_{wBT}	·	S_{wBT}	E_{DBT}	N_{pBT}	Q_{iBT}	W_{iBT}	t_{BT}	0	·	·	·
·	·	·	·	·	·	·	·	·	·	·	·	·
·	·	·	·	·	·	·	·	·	·	·	·	·
·	·	·	·	·	·	·	·	·	·	·	·	·
$1-S_{or}$	1.0	·	·	·	·	·	·	·	·	100%	0	·

Step 12. Express the results in a graphical form.

Chapter 3 Method to Calculate Sweep Efficiency

Section 1 Areal Sweep Efficiency

The areal sweep efficiency E_A is defined as the fraction of the total flood pattern that is contacted by the displacing fluid. It increases steadily with injection from zero at the start of the flood until breakthrough occurs, after which E_A continues to increase at a slower rate.

1. The Influence Factors of Areal Sweep Efficiency

The areal sweep efficiency depends basically on the following three main factors:

(1) Mobility ratio M;

(2) Flood pattern;

(3) Cumulative water injected W_{inj}.

1.1 Mobility ratio

In general, the mobility of any fluid λ is defined as the ratio of the effective permeability of the fluid to the fluid viscosity, i.e.:

$$\lambda_o = \frac{K_o}{\mu_o} = \frac{KK_{ro}}{\mu_o} \tag{3-1}$$

$$\lambda_w = \frac{K_w}{\mu_w} = \frac{KK_{rw}}{\mu_w} \tag{3-2}$$

$$\lambda_g = \frac{K_g}{\mu_g} = \frac{KK_{rg}}{\mu_g} \tag{3-3}$$

where λ_o, λ_w, λ_g——mobility of oil, water, and gas, respectively;

K_o, K_w, K_g——effective permeability to oil, water, and gas, respectively;

K_{ro}, K_{rw}——relative permeability to oil, water, and gas, respectively;

K——absolute permeability.

The fluid mobility as defined mathematically by the above three relationships indicates that λ is a strong function of the fluid saturation. The mobility ratio M is defined as the mobility of the displacing fluid to the mobility of the displaced fluid, or:

$$M = \frac{\lambda_{\text{displacing}}}{\lambda_{\text{displaced}}}$$

For water flooding, then:

$$M = \frac{\lambda_w}{\lambda_o}$$

Substituting for λ:

$$M = \frac{KK_{rw}}{\mu_w} \frac{\mu_o}{KK_{ro}}$$

Simplifying gives:

$$M = \frac{K_{rw}\mu_o}{K_{ro}\mu_w} \tag{3-4}$$

Muskat (1946) points out that in calculating M by applying Equation 3-4, the following concepts must be employed in determining K_{ro} and K_{rw}.

(1) Relative permeability of oil K_{ro}. Because the displaced oil is moving ahead of the water front in the noninvaded portion of the pattern, as shown schematically in Figure 3-1, K_{ro} must be evaluated at the initial water saturation S_{wi}.

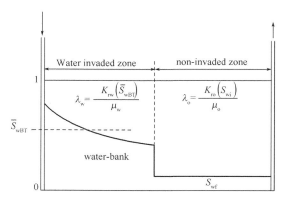

Figure 3-1 Oil and water mobilities to breakthrough

(2) Relative permeability of water K_{rw}. The displacing water will form a water bank that is characterized by an average water saturation of $\overline{S}_{wBT}$ in the swept area. This average saturation will remain constant until breakthrough, after which the average water saturation will continue to

increase (as denoted by $\bar{S}_{w2}$). The mobility ratio, therefore, can be expressed more explicitly under two different stages of the flood:

From the start to breakthrough:

$$M = \frac{K_{rw}(\bar{S}_{wBT})}{K_{ro}(S_{wi})} \frac{\mu_o}{\mu_w} \tag{3-5}$$

where $K_{rw}(\bar{S}_{wBT})$ ——relative permeability of water at $\bar{S}_{wBT}$;

$K_{ro}(S_{wi})$ ——relative permeability of oil at S_{wi}.

The above relationship indicates that the mobility ratio will remain constant from the start of the flood until breakthrough occurs.

After breakthrough:

$$M = \frac{K_{rw}(\bar{S}_{w2})}{K_{ro}(S_{wi})} \frac{\mu_o}{\mu_w} \tag{3-6}$$

Equation (3-6) indicates that the mobility of the water K_{rw}/μ_w will increase after breakthrough due to the continuous increase in the average water saturation $\bar{S}_{w2}$. This will result in a proportional increase in the mobility ratio M after breakthrough, as shown in Figure 3-2.

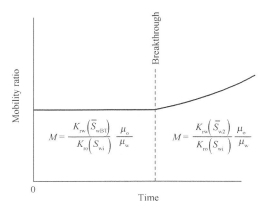

Figure 3-2 Mobility ratio versus time relationship

In general, if no further designation is applied, the term mobility ratio refers to the mobility ratio before breakthrough.

1.2　Flood patterns

In designing a waterflood project, it is common practice to locate injection and producing wells in a regular geometric pattern so that a symmetrical and interconnected network is formed. As shown previously, regular flood patterns include these:

(1) Direct line drive;

(2) Staggered line drive;

(3) Five spot;

(4) Seven spot;

(5) Nine spot.

By far the most used pattern is the five spot and, therefore, most of the discussion in the remainder of the chapter will focus on this pattern.

Craig et al. (1955) performed experimental studies on the influence of fluid mobilities on the areal sweep efficiency resulting from water or gas injection. Craig and his co-investigators used horizontal laboratory models representing a quadrant of five spot patterns. Areal sweep efficiencies were determined from X-ray shadowgraphs taken during various stages of the displacement as illustrated in Figure 3-3. Two mobility ratios, 1.43 and 0.4, were used in the study.

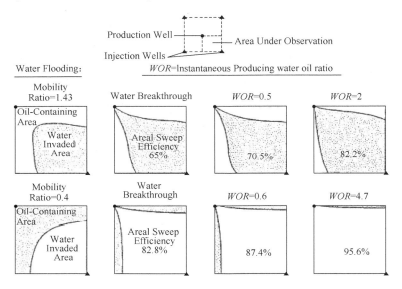

Figure 3-3　X-ray shadowgraphs of flood progress

Figure 3-3 shows that at the start of the flood, the water front takes on a cylindrical form around the injection point (well). As a result of the continuous injection, pressure distribution and corresponding streamlines are developed between the injection and production wells. However, various stream-lines have different lengths with the shortest streamline being the direct line between the injector and producer. The pressure gradient along this line is the highest that causes the injection fluid to flow faster along the shortest stream-line than the other lines. The water front gradually begins to deform from the cylindrical form and cusp into the production well as water breakthrough occurs. The effect of the mobility ratio on the areal sweep efficiency is apparent by examining Figure 3-3. This figure shows that at breakthrough, only 65% of the flood pattern area has been contacted (swept) by the injection fluid with a mobility ratio of 1.43 and 82.8% when the mobility ratio is 0.4. This contacted fraction when water breakthrough occurs is defined as the areal sweep efficiency at breakthrough, as denoted by E_{ABT}. In general, lower mobility ratios would increase the areal sweep efficiency and higher mobility ratios would decrease the E_A. Figure 14 – 34 also shows that with continued injection after breakthrough, the areal sweep efficiency continues to increase until it eventually reaches 100%.

1.3 Cumulative water injected

Continued injection after breakthrough can result in substantial increases in recovery, especially in the case of an adverse mobility ratio. The work of Craig et al. (1955) has shown that significant quantities of oil may be swept by water after breakthrough. It should be pointed out that the higher the mobility ratio, the more important is the "after-breakthrough" production.

2. Areal Sweep Prediction Methods

Methods of predicting the areal sweep efficiency are essentially divided into the following three phases of the flood:

(1) Before breakthrough;
(2) At breakthrough;
(3) After breakthrough.

2.1 Areal sweep efficiency before breakthrough

The areal sweep efficiency before breakthrough is simply proportional to the volume of water injected and is given by:

$$E_A = \frac{W_{inj}}{PV(\bar{S}_{wBT} - S_{wi})} \quad (3-7)$$

2.2 Areal sweep efficiency at breakthrough

Craig (1955) proposed a graphical relationship that correlates the areal sweep efficiency at breakthrough E_{ABT} with the mobility ratio for the five spot pattern. The correlation, as shown in Figure 3 – 4, closely simulates flooding operations and is probably the most representative of actual waterfloods. The graphical illustration of areal sweep efficiency as a strong function of mobility ratio shows that a change in the mobility ratio from 0.15 to 10.0 would change the breakthrough areal sweep efficiency from 100 to 50%. Willhite (1986) presented the following mathematical correlation, which closely approximates the graphical relationship presented in Figure 3 – 4.

$$E_{ABT} = 0.54602036 + \frac{0.03170817}{M} + \frac{0.30222997}{e^M} - 0.00509693M \quad (3-8)$$

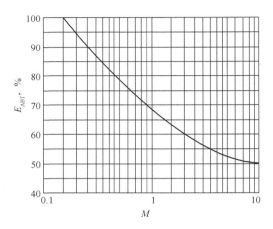

Figure 3-4 Areal sweep efficiency at breakthrough

2.3 Areal sweep efficiency after breakthrough

In the same way that displacement efficiency E_D increases after breakthrough, the areal sweep efficiency also increases due to the gradual increase in the total swept area with continuous injection. Dyes et al. (1954) correlated the increase in the areal sweep efficiency after breakthrough with the ratio of water volume injected at any time after breakthrough, W_{inj}, to water volume injected at breakthrough, W_{iBT}, as given by:

$$E_A = E_{ABT} + 0.633 \lg \frac{W_{inj}}{W_{iBT}} \quad (3-9)$$

or

$$E_A = E_{ABT} + 0.2749 \lg \frac{W_{inj}}{W_{iBT}} \quad (3-10)$$

Dyes et al. also presented a graphical relationship that relates the areal sweep efficiency with the reservoir water cut f_w and the reciprocal of mobility ratio $1/M$ as shown in Figure 3-5. Fassihi (1986) used a nonlinear regression model to reproduce the data of Figure 3-5 by using the following expression:

$$E_A = \frac{1}{1+A} \quad (3-11)$$

with

$$A = [a_1 \ln(M + a_2) + a_3] f_w + a_4 \ln(M + a_5) + a_6$$

Chapter 3 Method to Calculate Sweep Efficiency

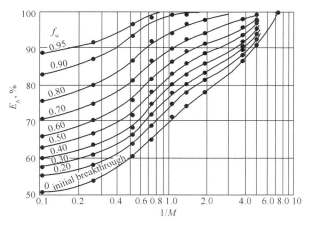

Figure 3 – 5 Areal sweep efficiency as a function of $1/M$ and f_w

The coefficient of Equation 3 – 11 for patterns such as the five spot, staggered line drive, and direct line drive are given in Table 3 – 1:

Table 3 – 1 Coefficients in areal sweep efficiency correlations

Coefficient	Five Spot	Direct Line	Staggered Line
a_1	– 0.2062	– 0.3014	– 0.2077
a_2	– 0.0712	– 0.1568	– 0.1059
a_3	– 0.511	– 0.9402	– 0.3526
a_4	0.3048	0.3714	0.2608
a_5	0.123	– 0.0865	0.2444
a_6	0.4394	0.8805	0.3158

Craig (1971) proposed that for a given value of E_{ABT} for a five-spot flood pattern, the ratio Q_i/Q_{iBT} that corresponds to W_{inj}/W_{iBT} could be determined mathematically by evaluating the following expression:

$$\frac{Q_i}{Q_{iBT}} = 1 + E_{ABT} \int_1^x \frac{1}{E_A} dx$$

with

$$x = \frac{W_{inj}}{W_{iBT}}; Q_i = 1/(df_w/dS_w)_{S_{w2}}, Q_{iBT} = 1/(df_w/dS_w)_{S_{wf}}$$

where Q_i——total pore volumes of water injected any time after breakthroug;

Q_{iBT}——total pore volumes of water injected at water breakthrough.

Craig tabulated the values of Q_i/Q_{iBT} as a function of W_{inj}/W_{iBT} and E_{ABT}. The author listed the values for a wide range of W_{inj}/W_{iBT} with E_{ABT} ranging from 50 to 90% as shown in Table 3 – 1. The value of Q_i/Q_{iBT} is read from the table for any particular value of E_{ABT} and the value of W_{inj}/W_{iBT} using interpolation if necessary. For example, if $E_{ABT}=70\%$ and $W_{inj}/W_{iBT}=2.00$, the value of the ratio Q_i/Q_{iBT} is read from Table 3-2 as 1.872, i.e., $Q_i/Q_{iBT}=1.872$.

Table 3-2 Q_i/Q_{iBT} values for various values of E_{ABT}

W_{inj}/W_{iBT}	$E_{ABT}(\%)$									
	50	51	52	53	54	55	56	57	58	59
1.0	1.000	1.000	1.000	1.000	1.000	1.000	1.000	1.000	1.000	1.000
1.2	1.190	1.191	1.191	1.191	1.191	1.191	1.191	1.191	1.192	1.192
1.4	1.365	1.366	1.366	1.367	1.368	1.368	1.369	1.369	1.370	1.370
1.6	1.529	1.530	1.531	1.532	1.533	1.535	1.536	1.536	1.537	1.538
1.8	1.684	1.686	1.688	1.689	1.691	1.693	1.694	1.696	1.697	1.699
2.0	1.832	1.834	1.837	1.839	1.842	1.844	1.846	1.849	1.851	1.853
2.2	1.974	1.977	1.981	1.984	1.987	1.990	1.993	1.996	1.999	2.001
2.4	2.111	2.115	2.119	2.124	2.127	2.131	2.135	2.139	2.142	2.146
2.6	2.244	2.249	2.254	2.259	2.264	2.268	2.273	2.277	2.282	2.286
2.8	2.373	2.379	2.385	2.391	2.397	2.402	2.407	2.413	2.418	2.422
3.0	2.500	2.507	2.513	2.520	2.526	2.533	2.539	2.545	2.551	2.556
3.2	2.623	2.631	2.639	2.646	2.653	2.660	2.667	2.674	2.681	2.687
3.4	2.744	2.752	2.761	2.770	2.778	2.786	2.793	2.801	2.808	2.816
3.6	2.862	2.872	2.881	2.891	2.900	2.909	2.917	2.926	2.934	2.942
3.8	2.978	2.989	3.000	3.010	3.020	3.030	3.039	3.048	3.057	3.066
4.0	3.093	3.105	3.116	3.127	3.138	3.149	3.159	3.169	3.179	3.189
4.2	3.205	3.218	3.231	3.243	3.254	3.266	3.277	3.288	3.299	3.309
4.4	3.316	3.330	3.343	3.357	3.369	3.382	3.394	3.406	3.417	3.428
4.6	3.426	3.441	3.455	3.469	3.483	3.496	3.509	3.521	3.534	3.546
4.8	3.534	3.550	3.565	3.580	3.594	3.609	3.622	3.636	3.649	
5.0	3.641	3.657	3.674	3.689	3.705	3.720	3.735			
5.2	3.746	3.764	3.781	3.798	3.814	3.830				
5.4	3.851	3.869	3.887	3.905	3.922					
5.6	3.954	3.973	3.993	4.011						
5.8	4.056	4.077	4.097							
6.0	4.157	4.179								
6.2	4.257									
Values of W_i/W_{iBT} at which $E_A=100$ percent										
	6.164	5.944	5.732	5.527	5.330	5.139	4.956	4.779	4.608	4.443

Chapter 3 Method to Calculate Sweep Efficiency

continued

$E_{ABT}(\%)$										
W_{inj}/W_{iBT}	60	61	62	63	64	65	66	67	68	69
1.0	1.000	1.000	1.000	1.000	1.000	1.000	1.000	1.000	1.000	1.000
1.2	1.192	1.192	1.192	1.192	1.192	1.192	1.193	1.193	1.193	1.193
1.4	1.371	1.371	1.371	1.372	1.372	1.373	1.373	1.373	1.374	1.374
1.6	1.539	1.540	1.541	1.542	1.543	1.543	1.544	1.545	1.546	1.546
1.8	1.700	1.702	1.703	1.704	1.706	1.707	1.708	1.709	1.710	1.711
2.0	1.855	1.857	1.859	1.861	1.862	1.864	1.866	1.868	1.869	1.871
2.2	2.004	2.007	2.009	2.012	2.014	2.16	2.019	2.021	2.023	2.025
2.4	2.149	2.152	2.155	2.158	2.161	2.164	2.167	2.170	2.173	2.175
2.6	2.290	2.294	2.298	2.301	2.305	2.308	2.312	2.315	2.319	2.322
2.8	2.427	2.432	2.436	2.441	2.445	2.449	2.453	2.457	2.461	2.465
3.0	2.562	2.567	2.572	2.577	2.582	2.587	2.592	2.597	2.601	2.606
3.2	2.693	2.700	2.705	2.711	2.717	2.723	2.728	2.733	2.738	2.744
3.4	2.823	2.830	2.836	2.843	2.849	2.855	2.862	2.867	2.873	
3.6	2.950	2.957	2.965	2.972	2.979	2.986	2.993			
3.8	3.075	3.083	3.091	3.099	3.107					
4.0	3.198	3.207	3.216	3.225						
4.2	3.319	3.329								
4.4	3.439									
Values of W_i/W_{iBT} at which $E_A = 100$ percent										
	4.235	4.132	3.984	3.842	3.704	3.572	3.444	3.321	3.203	3.088
$E_{ABT}(\%)$										
W_{inj}/W_{iBT}	70	71	72	73	74	75	76	77	78	79
1.0	1.000	1.000	1.000	1.000	1.000	1.000	1.000	1.000	1.000	1.000
1.2	1.193	1.193	1.193	1.193	1.193	1.193	1.193	1.194	1.194	1.194
1.4	1.374	1.375	1.375	1.375	1.376	1.376	1.376	1.377	1.377	1.377
1.6	1.547	1.548	1.548	1.549	1.550	1.550	1.551	1.551	1.552	1.552
1.8	1.713	1.714	1.715	1.716	1.717	1.718	1.719	1.720	1.720	1.721
2.0	1.872	1.874	1.875	1.877	1.878	1.880	1.881	1.882	1.884	1.885
2.2	2.027	2.029	2.031	2.033	2.035	2.037	2.039	2.040	2.042	2.044
2.4	2.178	2.180	2.183	2.185	2.188	2.190	2.192	2.195	2.197	
2.6	2.325	2.328	2.331	2.334	2.337	2.340				
2.8	2.469	2.473	2.476	2.480						
3.0	2.610	2.614								
Values of W_i/W_{iBT} at which $E_A = 100$ percent										
	2.978	2.872	2.769	2.670	2.575	2.483	2.394	2.309	2.226	2.147

continued

W_{inj}/W_{iBT}	$E_{ABT}(\%)$									
	80	81	82	83	84	85	86	87	88	89
1.0	1.000	1.000	1.000	1.000	1.000	1.000	1.000	1.000	1.000	1.000
1.2	1.194	1.194	1.194	1.194	1.194	1.194	1.194	1.194	1.194	1.194
1.4	1.377	1.378	1.378	1.378	1.378	1.379	1.379	1.379	1.379	1.379
1.6	1.553	1.553	1.554	1.555	1.555	1.555	1.556	1.556	1.557	1.557
1.8	1.722	1.723	1.724	1.725	1.725	1.726	1.727	1.728		
2.0	1.886	1.887	1.888	1.890						
2.2	2.045									
Values of W_i/W_{iBT} at which E_A = 100 percent										
	2.070	1.996	1.925	1.856	1.790	1.726	1.664	1.605	1.547	1.492
W_{inj}/W_{iBT}	$E_{ABT}(\%)$									
	90	91	92	93	94	95	96	97	98	99
1.0	1.000	1.000	1.000	1.000	1.000	1.000	1.000	1.000	1.000	1.000
1.2	1.194	1.195	1.195	1.195	1.195	1.195	1.195	1.195	1.195	1.195
1.4	1.380	1.380	1.380	1.380	1.381					
1.6	1.558									
Values of W_i/W_{iBT} at which E_A = 100 percent										
	1.439	1.387	1.338	1.290	1.244	1.199	1.157	1.115	1.075	1.037

Willhite (1986) proposed an analytical expression for determining the value of the ratio (Q_i/Q_{iBT}) at any value of (W_{inj}/W_{iBT}) for a given E_{ABT}:

$$\frac{Q_i}{Q_{iBT}} = 1 + a_1 e^{-a_1} [\text{Ei}(a_2) - \text{Ei}(a_1)] \qquad (3-12)$$

where,

$$a_1 = 3.65 E_{ABT}$$

$$a_2 = a_1 + \ln \frac{W_{inj}}{W_{iBT}}$$

and Ei(x) is the Ei function as approximated by:

$$\text{Ei}(x) = 0.57721557 + \ln(x) + \sum_{n=1}^{\infty} \frac{x^n}{n(n!)}$$

3. Including the Areal Sweep Efficiency in Waterflooding Calculation

To include the areal sweep efficiency in waterflooding calculations, the proposed methodology is divided into the following three phases:

(1) Initial calculations;

(2) Recovery performance calculations to breakthrough;

(3) Recovery performance calculations after breakthrough.

The specific steps of each of the above three phases are summarized below.

3.1 Initial calculations ($S_{gi} = 0$, $E_V = 100\%$)

Step 1. Express the relative permeability data as relative permeability ratios and plot them versus their corresponding water saturations on a semi-log scale. Describe the resulting straight line by the following relationship:

$$\frac{K_{ro}}{K_{rw}} = a e^{bS_w}$$

Step 2. Calculate and plot f_w versus S_w.

Step 3. Draw a tangent to the fractional flow curve as originated from S_{wi} and determine:

(1) Point of tangency (S_{wf}, f_{wf}), i.e., (S_{wBT}, f_{wBT});

(2) Average water saturation at breakthrough $\bar{S}_{wBT}$;

(3) Slope of the tangent $(df_w/dS_w)_{S_{wf}}$.

Step 4. Using S_{wi} and $\bar{S}_{wBT}$, determine the corresponding values of K_{ro} and K_{rw}. Designate these values $K_{ro}(S_{wi})$ and $K_{rw}(\bar{S}_{wBT})$, respectively.

Step 5. Calculate the mobility ratio:

$$M = \frac{K_{rw}(\bar{S}_{wBT})}{K_{ro}(S_{wi})} \frac{\mu_o}{\mu_w}$$

Step 6. Select several water saturations S_{w2} between S_{wf} and $(1 - S_{or})$ and numerically or graphically determine the slope $(df_w/dS_w)_{S_{w2}}$ at each saturation.

Step 7. Plot $(df_w/dS_w)_{S_{w2}}$ versus S_{w2} on a Cartesian scale.

3.2 Recovery performance to breakthrough

Assuming that the vertical sweep efficiency E_V and initial gas saturation S_{gi} are 100% and 0%, respectively, the required steps to complete the calculations of this phase are summarized below.

Step 1. Calculate the areal sweep efficiency at breakthrough E_{ABT}.

Step 2. Calculate pore volumes of water injected at breakthrough:

$$Q_{iBT} = \frac{1}{\left(\dfrac{df_w}{dS_w}\right)_{S_{wf}}} = (\bar{S}_{wBT} - S_{wi})$$

Step 3. Calculate cumulative water injected at breakthrough W_{iBT}.

Step 4. Assuming a constant water-injection rate i_w, calculate time to breakthrough t_{BT}:

$$t_{BT} = \frac{W_{iBT}}{i_w}$$

Step 5. Calculate the displacement efficiency at breakthrough E_{DBT}:

$$E_{DBT} = \frac{\overline{S}_{wBT} - S_{wi}}{1 - S_{wi}}$$

Step 6. Compute the cumulative oil production at breakthrough:

$$(N_p)_{BT} = N_S E_{DBT} E_{ABT}$$

Notice that when $S_{gi} = 0$, the cumulative oil produced at breakthrough is equal to cumulative water injected at breakthrough, or:

$$(N_p)_{BT} = \frac{W_{iBT}}{B_o}$$

Step 7. Divide the interval between 0 and W_{iBT} into any arbitrary number of increments and set the following production data for each increment:

$$Q_o = i_w / B_o$$

$$Q_w = 0$$

$$WOR = 0$$

$$N_p = W_{inj}/B_o$$

$$W_p = 0$$

$$t = W_{inj}/i_w$$

Step 8. Express steps 1 through 7 in Table 3-3.

Table 3-3 The preceding calculations

W_{inj}	$t = W_{inj}/i_w$	$N_p = W_{inj}/B_o$	$Q_o = i_w/B_o$	WOR_s	$Q_w = Q_o WOR_s$	W_p
0	0	0	0	0	0	0
.	.	.	.	0	0	0
.	.	.	.	0	0	0
.	.	.	.	0	0	0
W_{iBT}	t_{BT}	$(N_p)_{BT}$	.	WOR_s	.	0

3.3 Recovery performance after breakthrough ($S_{gi} = 0$, $E_V = 100\%$)

Craig et al. (1955) point out that after water breakthrough, the displacing fluid continues to displace more oil from the already swept zone (behind the front) and from newly swept regions in the pattern. Therefore, the producing water-oil ratio WOR_s is estimated by separating the

Chapter 3　Method to Calculate Sweep Efficiency

displaced area into two distinct zones:

(1) Previously swept area of the flood pattern;

(2) Newly swept zone that is defined as the region that was just swept by the displacing fluid.

The previously swept area contains all reservoir regions where water saturation is greater than S_{wf} and continues to produce both oil and water. With continuous water injection, the injected water contacts more regions as the area sweep efficiency increases. This newly swept zone is assumed to produce only oil. Craig et al. (1955) developed an approach for determining the producing WOR_s that is based on estimating the incremental oil produced, $(\Delta N_p)_{newly}$, from the newly swept region for 1 bbl of total production. The incremental oil produced from the newly swept zone is given by:

$$(\Delta N_p)_{newly} = E\lambda(Q_{iBT})E_{ABT} \qquad (3-13)$$

with

$$E = \frac{S_{wf} - S_{wi}}{E_{ABT}(\overline{S}_{wBT} - S_{wi})}, \lambda = 0.2749\frac{W_{iBT}}{W_{inj}}$$

Notice that the parameter E is constant, whereas the parameter is decreasing with continuous water injection. Craig et al. expressed the producing water-oil ratio as:

$$WOR_s = \frac{f_{w2}(1-(\Delta N_p)_{newly})}{1-[f_{w2}(1-(\Delta N_p)_{newly})]}\frac{B_o}{B_w} \qquad (3-14)$$

Note that when the areal sweep efficiency E_A reaches 100%, the incremental oil produced from the newly swept areal is zero, i.e., $(N_p)_{newly} = 0$, which reduces the above expression to Equation:

$$WOR_s = \frac{f_{w2}}{1-f_{w2}}\frac{B_o}{B_w} = \frac{B_o}{B_w\left(\frac{1}{f_{w2}}-1\right)}$$

The recommended methodology for predicting the recovery performance after breakthrough is summarized in the following steps:

Step 1. Select several values of $W_{inj} > W_{iBT}$.

Step 2. Assuming constant injection rate i_w, calculate the time t required to inject W_{inj} barrels of water.

Step 3. Calculate the ratio W_{inj}/W_{iBT} for each selected W_{inj}.

Step 4. Calculate the areal sweep efficiency E_A at each selected W_{inj}:

$$E_A = E_{ABT} + 0.6331\lg\frac{W_{inj}}{W_{iBT}} = E_{ABT} + 0.2749\ln\frac{W_{inj}}{W_{iBT}}$$

Step 5. Calculate the ratio Q_i/Q_{iBT} that corresponds to each W_{inj}/W_{iBT}. The ratio Q_i/Q_{iBT} is a function of E_{ABT} and W_{inj}/W_{iBT}.

Step 6. Determine the total pore volumes of water injected by multiplying each ratio of Q_i/Q_{iBT} (obtained in step 5) by Q_{iBT}, or:

$$Q_i = \frac{Q_i}{Q_{iBT}} Q_{iBT}$$

Step 7. From the definition of Q_i, determine the slope $(df_w/dS_w)_{S_{w2}}$ for each value of Q_i by:

$$\left(\frac{df_w}{dS_w}\right)_{S_{w2}} = \frac{1}{Q_i}$$

Step 8. Read the value of S_{w2}, i.e., water saturation at the producing well, that corresponds to each slope from the plot of $(df_w/dS_w)_{S_{w2}}$ vs. S_{w2} (see phase 1, step 7).

Step 9. Calculate the reservoir water cut at the producing well f_{w2} for each S_{w2}:

$$f_{w2} = \frac{1}{1 + \frac{\mu_w}{\mu_o}\frac{K_{ro}}{K_{rw}}}$$

or

$$f_{w2} = \frac{1}{1 + \frac{\mu_w}{\mu_o}ae^{bS_{w2}}}$$

Step 10. Determine the average water saturation in the swept area $\overline{S}_{w2}$:

$$\overline{S}_{w2} = S_{w2} + \frac{1 - f_{w2}}{\left(\dfrac{df_w}{dS_w}\right)_{S_{w2}}}$$

Step 11. Calculate the displacement efficiency E_D for each S_{w2}:

$$E_D = \frac{\overline{S}_{w2} - S_{wi}}{1 - S_{wi}}$$

Step 12. Calculate cumulative oil production:

$$N_p = N_S E_D E_A E_V$$

for 100% vertical sweep efficiency:

$$N_p = N_S E_D E_A$$

Step 13. Calculate cumulative water production:

$$W_p = \frac{W_{inj} - N_p B_o}{B_w}$$

$$W_p = \frac{W_{inj} - \left(\overline{S}_{w2} - S_{wi}\right) PVE_A}{B_w}$$

Step 14. Calculate the surface water-oil ratio WOR_s that corresponds to each value of f_{w2}:

$$WOR_s = \frac{f_{w2}[1-(\Delta N_p)_{newly}]}{1-[f_{w2}(1-(\Delta N_p)_{newly})]} \frac{B_o}{B_w}$$

Step 15. Calculate the oil and water flow rates, respectively:

$$Q_o = \frac{i_w}{B_o + B_w WOR_s}$$

$$Q_w = Q_o WOR_s$$

Steps 1 through 15 could be conveniently performed in the following worksheet form (Table 3-4).

Table 3-4 The preceding calculations

W_{inj}	$t = \frac{W_{inj}}{i_w}$	$\frac{W_{inj}}{W_{iBt}}$	E_A	$\frac{Q_i}{Q_{iBT}}$	Q_i	$\left(\frac{df_w}{dS_w}\right)_{S_w}$	S_{w2}	f_{w2}	$\bar{S}_{w2}$	E_D	N_p	W_p	WOR_s	Q_o	Q_w
W_{iBT}	t_{BT}	1.0	E_{ABT}	1.0	Q_{iBT}	—	S_{wBT}	f_{wBT}		E_{DBT}	—	—	—	—	—

Note that all the areal sweep efficiency correlations that have been presented thus far are based on idealized cases with severe imposed assumptions on the physical characteristics of the reservoir. These assumptions include:

(1) Uniform isotropic permeability distribution;

(2) Uniform porosity distribution;

(3) No fractures in reservoir;

(4) Confined patterns;

(5) Uniform saturation distribution;

(6) Off-pattern wells.

To understand the effect of eliminating any of the above assumptions on the areal sweep efficiency, it has been customary to employ laboratory models to obtain more generalized numerical expressions. However, it is virtually impossible to develop a generalized solution when eliminating all or some of the above assumptions.

Landrum and Crawford (1960) have studied the effects of directional permeability on waterflood areal sweep efficiency. Figure 3-6 and Figure 3-7 illustrate the impact of directional permeability variations on areal sweep efficiency for a line drive and five-spot pattern flood.

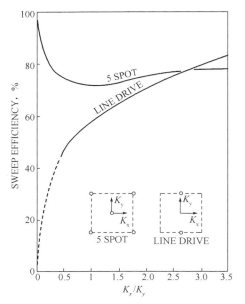
Figure 3-6 Effect of directional permeability on E_A

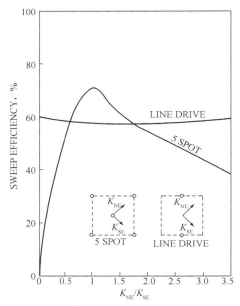
Figure 3-7 Effect of directional permeability on E_A

4. Elements Affect the Performance of Waterflooding

Three key elements affect the performance of waterflooding that must be included in recovery calculations: (1) Water injection rate, i.e., fluid injectivity; (2) Effect of initial gas saturation on the recovery performance; (3) Water fingering and tonguing.

4.1 Fluid injectivity

Injection rate is a key economic variable that must be considered when evaluating a waterflooding project. The waterflood project's life and, consequently, the economic benefits will be directly affected by the rate at which fluid can be injected and produced. Estimating the injection rate is also important for the proper sizing of injection equipment and pumps. Although injectivity can be best determined from small-scale pilot floods, empirical methods for estimating water injectivity for regular pattern floods have been proposed by Muskat (1948) and Deppe. The authors derived their correlations based on the following assumptions:

(1) Steady-state conditions;
(2) No initial gas saturation;
(3) Mobility ratio of unity.

Water injectivity is defined as the ratio of the water injection to the pressure difference between the injector and producer, or:

Chapter 3　Method to Calculate Sweep Efficiency

$$I = \frac{i_w}{\Delta P}$$

where　I——injectivity, bbl/(d · psi);

　　　i_w——injection rate, bbl/d;

　　　ΔP——difference between injection pressure and producing well bottom hole flowing pressure.

When the injection fluid has the same mobility as the reservoir oil (mobility ratio $M = 1$), the initial injectivity at the start of the flood is referred to as I_{base}, or:

$$I_{base} = \frac{i_{base}}{\Delta P_{base}}$$

where　i_{base}——initial (base) water injection rate, bbl/d;

　　　ΔP_{base}——initial (base) pressure difference between injector and producer.

For a five-spot pattern that is completely filled with oil, i.e., $S_{gi} = 0$, Muskat proposed the following injectivity equation:

$$I_{base} = \frac{0.003541 h K K_{ro} \Delta P_{base}}{\mu_o \left(\ln \frac{d}{r_w} - 0.619 \right)} \tag{3-15}$$

or

$$\left(\frac{i}{\Delta P} \right)_{base} = \frac{0.003541 h K K_{ro}}{\mu_o \left[\ln \frac{d}{r_w} - 0.619 \right]} \tag{3-16}$$

where　i_{base}——base (initial) water injection rate, bbl/d;

　　　h——net thickness, ft;

　　　K——absolute permeability, mD;

　　　K_{ro}——oil relative permeability as evaluated at S_{wi};

　　　ΔP_{base}——base (initial) pressure difference, psi;

　　　d——distance between injector and producer, ft;

　　　r_w——wellbore radius, ft.

Several studies have been conducted to determine the fluid injectivity at mobility ratios other than unity. All of the studies concluded the following:

(1) At favorable mobility ratios, i.e., $M < 1$, the fluid injectivity declines as the areal sweep efficiency increases.

(2) At unfavorable mobility ratios, i.e., $M > 1$, the fluid injectivity increases with

increasing areal sweep efficiency.

Caudle and Witte (1959) used the results of their investigation to develop a mathematical expression that correlates the fluid injectivity with the mobility ratio and areal sweep efficiency for five – spot patterns. The correlation may only be used in a liquid-filled system, i. e. , $S_{gi}=0$. The authors presented their correlation in terms of the conductance ratio γ, which is defined as the ratio of the fluid injectivity at any stage of the flood to the initial (base) injectivity, i. e. :

$$\gamma = \frac{\dfrac{i_w}{\Delta P}}{\left(\dfrac{i}{\Delta P}\right)_{base}} \qquad (3-17)$$

Caudle and Witte presented the variation in the conductance ratio with E_A and M in graphical form as shown in Figure 3 – 8. Note again that if an initial gas is present, the Caudle-Witte conductance ratio will not be applicable until the gas is completely dissolved or the system becomes liquid filled (fill-up occurs). The two possible scenarios for the practical use of Equation (3 – 17) follow:

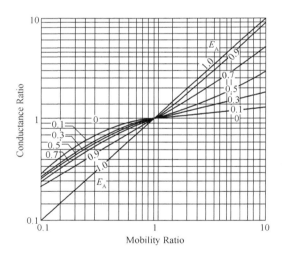

Figure 3 – 8 Conductance ratio curve

Scenario 1: Constant Injection Pressure and Variable Injection Rate

At constant injection pressure, i. e. , $\Delta P_{base} = \Delta P$, the conductance ratio as expressed by Equation can be written as:

$$\gamma = \frac{i_w}{i_{base}}$$

or

$$i_w = \gamma i_{base} \quad (3-18)$$

Scenario 2: Constant Injection Rate and Variable Injection Pressure

When the water injection rate is considered constant, i.e., $i_w = i_{base}$, the conductive ratio is expressed as:

$$\gamma = \frac{\Delta P_{base}}{\Delta P}$$

or

$$\Delta P = \frac{\Delta P_{base}}{\gamma} \quad (3-19)$$

4.2 Effect of initial gas saturation

When a solution-gas-drive reservoir is under consideration for water-flooding, substantial gas saturation usually exists in the reservoir at the start of the flood. It is necessary to inject a volume of water that approaches the volume of the pore space occupied by the free gas before the oil is produced. This volume of water is called the fill-up volume. Because economic considerations dictate that waterflooding should occur at the highest possible injection rates, the associated increase in the reservoir pressure might be sufficient to redissolve all of the trapped gas Sgt back in solution. Willhite (1986) points out that relatively small increases in pressure frequently are required to redissolve the trapped gas. Thus in waterflooding calculations, it is usually assumed that the trapped (residual) gas saturation is zero. A description of the displacement mechanism occurring under a five-spot pattern will indicate the nature of other secondary recovery operations. The five-spot pattern uses a producing well and four injection wells. The four injectors drive the crude oil inward to the centrally located producer. If only one five-spot pattern exists, the ratio of injection to producing wells is 4:1; however, on a full-field scale it includes a large number of adjacent five spots. In such a case, the number of injection wells compared to producing wells approaches a 1:1 ratio.

At the start of the waterflood process in a solution-gas-drive reservoir, the selected flood pattern is usually characterized by a high initial gas saturation of S_{gi} and remaining liquid saturations of S_{oi} and S_{wi}. When initial gas saturation exists in the reservoir, Craig, Geffen, and Morse (1955) developed a methodology that is based on dividing the flood performance into four stages. The method, known as the CGM method after the authors, was developed from experimental data in horizontal laboratory models representing a quadrant of a five spot. Craig et al. identified the following four stages of the waterflood as:

(1) Start-interference;

(2) Interference-fill-up;

(3) Fill-up-water breakthrough;

(4) Water breakthrough-end of the project.

A detailed description of each stage of the flood is illustrated schematically in figures and described below:

1) Start-interference

At the start of the water-injection process in the selected pattern area of a solution-gas-drive reservoir, high gas saturation usually exists in the flood area as shown schematically in Figure 3-9. The current oil production at the start of the flood is represented by point A on the conventional flow rate-time curve of Figure 3-10. After the injection is initiated and a certain amount of water injected, an area of high water saturation called the water bank is formed around the injection well at the start of the flood. This stage of the injection is characterized by a radial flow system for both the displacing water and displaced oil. With continuous water injection, the water bank grows radially and displaces the oil phase that forms a region of high oil saturation that forms an oil bank. This radial flow continues until the oil banks, formed around adjacent injectors, meet. The place where adjacent oil banks meet is termed Interference, as shown schematically in Figure 3-11. During this stage of the flood, the condition around the producer is similar to that of the beginning of the flood, i.e., no changes are seen in the well flow rate Q_o as indicated in Figure 3-10 by point B. Craig, Geffen, and Morse (1955) summarized the computational steps during this stage of the flood, where radial flow prevails, in the following manner.

Step 1. Calculate the cumulative water injected to interference W_{ii} from the following expression:

$$W_{ii} = \frac{\pi h \phi S_{gi} r_{ei}^2}{5.615} \tag{3-20}$$

where W_{ii}——cumulative water injected to interference, bbl;

S_{gi}——initial gas saturation;

ϕ——porosity;

r_{ei}——half the distance between adjacent injectors, ft.

Step 2. Assume several successive values of cumulative water injected W_{inj}, ranging between 0 and W_{ii}, and calculate the water-injection rate at each assumed value of W_{inj} from:

Chapter 3 Method to Calculate Sweep Efficiency

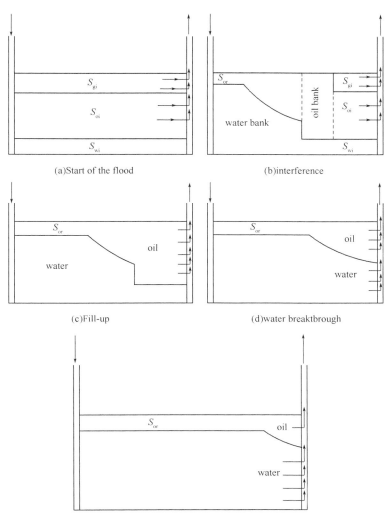

Figure 3 – 9 Stages of waterflooding

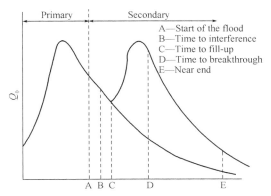

Figure 3 – 10 Predicted production history

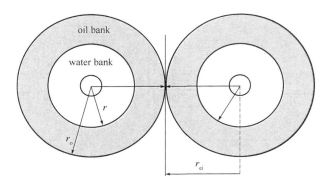

Figure 3-11 Interference of oil banks

$$i_w = \frac{0.00707hk\Delta p}{\left(\dfrac{\mu_w}{k_{rw}}\ln\dfrac{r}{r_w} + \dfrac{\mu_o}{k_{ro}}\ln\dfrac{r_o}{r}\right)} \tag{3-21}$$

where r_o——outer radius of the oil bank, ft;

r——outer radius of the water bank, ft;

r_w——wellbore radius, ft.

The outer radii of the oil and water banks are calculated from:

$$r_o = \sqrt{\frac{5.615 W_{inj}}{\pi h \phi S_{gi}}} \tag{3-22}$$

$$r = r_o \sqrt{\frac{S_{gi}}{S_{wBT} - S_{wi}}} \tag{3-23}$$

2) Interference to fill-up

This stage describes the period from interference until the fill-up of the preexisting gas space. Fill-up is the start of oil production response as illustrated in Figure 3-9 and by point C on Figure 3-10. The flow during this time is not strictly radial and is generally complex to quantify mathematically. Therefore, the flood performance can only be determined at the time of fill-up.

The required performance calculations at the fill-up are summarized in the following steps:

Step 1. Calculate the cumulative water injected at fill-up W_{if} by applying the following expression:

$$W_{if} = PV S_{gi} \tag{3-24}$$

The above equation suggests that while fill-up is occurring, the oil production rate is either zero or negligible, compared with the water injection rate. If the oil production rate Q_o prior to fill-up is significant, the cumulative water injected at the fill-up W_{if} must be increased by the total volume of oil produced from the start of injection to fill-up, i.e.:

$$W_{if} = PVS_{gi} + \frac{N_p}{B_o} \qquad (3-25)$$

where N_p——cumulative oil production from start of flood to fill-up, bbl;

B_o——oil formation volume factor, bbl/bbl.

Equation(3 – 25) indicates that the fill-up time will also increase, in addition, it causes the fill-up time calculation to be iterative.

Step 2. Calculate the areal sweep efficiency at fill-up

$$E_A = \frac{W_{inj}}{PV(\bar{S}_{wBT} - S_{wi})}$$

at fill-up:

$$E_A = \frac{W_{if}}{PV(\bar{S}_{wBT} - S_{wi})}$$

Step 3. Using the mobility ratio and the areal sweep efficiency at fill-up, determine the conductance ratio γ. Note that the conductance ratio can only be determined when the flood pattern is completely filled with liquids, which occurs at the fill-up stage.

Step 4. For a constant pressure difference, the initial (base) water injection rate i_{base} is:

$$i_{base} = \frac{0.003541 h K K_{ro} \Delta P}{\mu_o \left(\ln \frac{d}{r_w} - 0.619 \right)}$$

Step 5. Calculate the water injection at fill-up i_{wf}:

$$i_{wf} = \gamma i_{base}$$

The above expression is only valid when the system is filled with liquid, i.e., from the fill-up point and thereafter.

Step 6. Calculate the incremental time occurring from interference to fill-up from:

$$\Delta t = \frac{W_{if} - W_{ii}}{\dfrac{i_{wi} + i_{wf}}{2}}$$

The above expression suggests that the fill-up will occur after interference.

3) Fill-up to water breakthrough

The time to fill-up, as represented by point C on Figure 3 – 10, marks the following four events:

(1) No free gas remaining in the flood pattern;

(2) Arrival of the oil-bank front to the production well;

(3) Flood pattern response to the waterflooding;

(4) Oil flow rate Q_o equal to the water injection rate i_w.

During this stage, the oil production rate is essentially equal to the injection due to the fact that no free gas exists in the swept flood area. With continuous water injection, the leading edge of the water bank eventually reaches the production well, as shown in Figure 3-9, and marks the time to breakthrough. At breakthrough the water production rises rapidly.

The waterflood performance calculations are given by the following steps:

Step 1. Calculate cumulative water injected at breakthrough:

$$W_{iBT} = PV(\bar{S}_{wBT} - S_{wi})E_{ABT} = PVQ_{iBT}E_{ABT}$$

Step 2. Assume several values of cumulative water injected W_{inj} between W_{if} and W_{iBT} and calculate the areal sweep efficiency at each W_{inj}:

$$E_A = \frac{W_{inj}}{PV(\bar{S}_{wBT} - S_{wi})}$$

Step 3. Determine the conductance ratio γ for each assumed value of W_{inj}.

Step 4. Calculate the water injection rate at each W_{inj}:

$$i_w = \gamma i_{base}$$

Step 5. Calculate the oil flow rate Q_o during this stage from:

$$Q_o = \frac{i_w}{B_o}$$

Step 6. Calculate cumulating oil production N_P from the following expression:

$$N_p = \frac{W_{inj} - W_{if}}{B_o}$$

4) Water breakthrough to end of the project

After breakthrough, the water-oil ratio increases rapidly with a noticeable decline in the oil flow rate as shown in Figure 3-10 by point D. The swept area will continue to increase as additional water is injected. The incrementally swept area will contribute additional oil production, while the previously swept area will continue to produce both oil and water.

As represented by Equation, the WOR is calculated on the basis of the amounts of oil and water flowing from the swept region and the oil displaced from the newly swept portion of the pattern. It is assumed the oil from the newly swept area is displaced by the water saturation just behind the stabilized zone, i.e., S_{wf}.

The calculations during the fourth stage of the waterflooding process are given below:

Step 1. Assume several values for the ratio W_{inj}/W_{iBT} that correspond to the values given in Table, i.e., 1, 1.2, 1.4, etc.

Chapter 3 Method to Calculate Sweep Efficiency

Step 2. Calculate the cumulative water injected for each assumed ratio of W_{inj}/W_{iBT} from:

$$W_{inj} = \frac{W_{inj}}{W_{iBT}} W_{iBT}$$

Step 3. Calculate the areal sweep efficiency at each assumed W_{inj}/W_{iBT}:

$$E_A = E_{ABT} + 0.6331 \lg \frac{W_{inj}}{W_{iBT}}$$

Step 4. Calculate the ratio Q_i/Q_{iBT} that corresponds to each value of W_{inj}/W_{iBT} from Table.

Step 5. Determine the total pore volumes of water injected by multiplying each ratio of Q_i/Q_{iBT} by Q_{iBT}, or:

$$Q_i = \frac{Q_i}{Q_{iBT}} Q_{iBT}$$

Step 6. From the definition of Q_i, determine the slope $(df_w/dS_w)_{S_{w2}}$ for each value of Q_i by:

$$\left(\frac{df_w}{dS_w}\right)_{S_{w2}} = \frac{1}{Q_i}$$

Step 7. Read the value of S_{w2}, i.e., water saturation at the producing well, that corresponds to each slope from the plot of $(df_w/dS_w)S_{w2}$ vs. S_{w2}.

Step 8. Calculate the reservoir water cut at the producing well f_{w2} for each S_{w2}:

$$f_{w2} = \frac{1}{1 + \dfrac{\mu_w K_{ro}}{\mu_o K_{rw}}}$$

or

$$f_{w2} = \frac{1}{1 + \left(\dfrac{\mu_w}{\mu_o}\right) a e^{bS_{w2}}}$$

Step 9. Determine the average water saturation in the swept area $\bar{S}_{w2}$:

$$\bar{S}_{w2} = S_{w2} + \frac{1 - f_{w2}}{\left(\dfrac{df_w}{dS_w}\right)_{S_{w2}}}$$

Step 10. Calculate the surface water-oil ratio WOR_s that corresponds to each value of f_{w2}:

$$WOR_s = \frac{f_{w2}[1 - (\Delta N_p)_{newly}] B_o}{1 - f_{w2}[1 - (\Delta N_p)_{newly}] B_w}$$

Step 11. Craig, Geffen, and Morse (1955) point out when calculating cumulative oil production during this stage that one must account for the oil lost to the unswept area of the flood pattern. To account for the lost oil, the authors proposed the following expression:

$$N_p = N_S E_D E_A - \frac{PV(1-E_A)S_{gi}}{B_o}$$

where, E_D is the displacement efficiency and is given as:

$$E_D = \frac{\overline{S}_w - S_{wi} - S_{gi}}{1 - S_{wi} - S_{gi}}$$

Step 12. Calculate cumulative water from the expression:

Water produced = Water injected Oil produced Fill up volume

or

$$W_p = \frac{W_{inj} - N_p B_o - (PV)S_{gi}}{B_w}$$

Step 13. Calculate K_{rw} at $\overline{S}_{w2}$ and determine the mobility ratio M after breakthrough:

$$M = \frac{K_{rw}(\overline{S}_{w2})}{K_{ro}(S_{wi})} \frac{\mu_o}{\mu_w}$$

Step 14. Calculate the conductance ratio.

Step 15. Determine the water injection rate:

$$i_w = \gamma i_{base}$$

Step 16. Calculate the oil and water production rates, respectively:

$$Q_o = \frac{i_w}{B_o + B_w WOR_s}$$

$$Q_w = Q_o WOR_s$$

4.3 Water fingering and tonguing

In thick, dipping formations containing heavy viscous oil, water tends to advance as a "tongue" at the bottom of the pay zone. Similarly, displacement of oil with a gas will result in the gas attempting to overrun the oil due to gravity differences unless stopped by a shale barrier within the formation or by a low overall effective vertical permeability. In linear laboratory experiments, it was observed that the fluid interface remains horizontal and independent of fluid velocity when the viscosities of the two phases are equal. If the oil and water have different viscosities, the original horizontal interface will become tilted.

In a dipping reservoir, Dake (1978) developed a gravity segregation model that allows the calculation of the critical injection rate i_{crit} that is required to propagate a stable displacement. The condition for stable displacement is that the angle between the fluid interface and the direction of flow should remain constant throughout the displacement as shown in Figure3 – 12. Dake introduced the two parameters, the Dimensionless Gravity Number "G" and the End – point

Mobility Ratio M^*, that can be used to define the stability of displacement. These two parameters are defined by the following relationships.

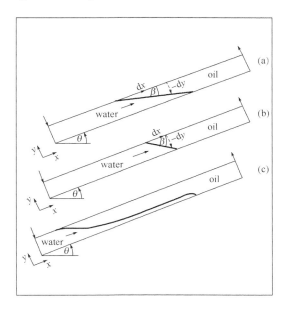

Figure 3-12 Stable and unstable displacement in gravity segregated displacement
(a) stable: $G > M^* - 1, M^* > 1$, and $\beta < \theta$; (b) stable: $G > M^* - 1, M^* < 1, \beta > \theta$; and
(c) unstable: $G < M^* - 1$

(1) Dimensionless gravity number. The dimensionless gravity number G is given by:

$$G = \frac{7.853 \times 10^{-6} K K_{rw} A (\rho_w - \rho_o) \sin\theta}{i_w \mu_w} \quad (3-26)$$

where K——absolute permeability, mD;

K_{rw}——relative permeability to water as evaluated at S_{or};

A——cross-sectional area;

ρ_w——water density, lb/ft^3;

θ——dip angle, (°).

(2) End-point mobility ratio. The end-point mobility ratio M^* is defined by:

$$M^* = \frac{K_{rw}(S_{or})\mu_o}{K_{ro}(S_{wi})\mu_o} \quad (3-27)$$

Dake used the above two parameters to define the following stability criteria:

(1) If $M^* > 1$. The displacement is stable if $G > (M^* - 1)$, in which case the fluid interface angle $< \theta$. The displacement is unstable if $G < (M^* - 1)$.

(2) If $M^* = 1$. This is a very favorable condition, because there is no tendency for the water

to by pass the oil. The displacement is considered unconditionally stable and is characterized by the fact that the interface rises horizontally in the reservoir, i. e. , $\beta = \theta$.

(3) If $M^* < 1$. When the end-point mobility ratio M^* is less than unity, the displacement is characterized as unconditionally stable displacement with $\beta > \theta$.

Dake also defined the critical flow rate, i_{crit} by:

$$i_{crit} = \frac{7.853 \times 10^{-6} KK_{rw} A (\rho_w - \rho_o) \sin\theta}{\mu_w (M^* - 1)} \quad (3-28)$$

Dake (1978) and Willhite (1986) presented a comprehensive treatment of water flooding under segregated flow conditions.

Section 2 Vertical Sweep Efficiency

The vertical sweep efficiency, E_V, is defined as the fraction of the vertical section of the pay zone that is swept by the injection fluid. This particular sweep efficiency depends primarily on (1) the mobility ratio and (2) total volume injected. As a consequence of the nonuniform permeabilities, any injected fluid will tend to move through the reservoir with an irregular front. In the more permeable portions, the injected water will travel more rapidly than in the less permeable zone.

1. The Nonuniform Permeability

Perhaps the area of the greatest uncertainty in designing a waterflood is the quantitative knowledge of the permeability variation within the reservoir. The degree of permeability variation is considered by far the most significant parameter influencing the vertical sweep efficiency.

To calculate the vertical sweep efficiency, the engineer must be able to address the following three problems:

(1) How to describe and define the permeability variation in mathematical terms.

(2) How to determine the minimum number of layers that are sufficient to model the performance of the fluid.

(3) How to assign the proper average rock properties for each layer (called the zonation problem).

A complete discussion of the above three problems is given below.

1.1 Reservoir vertical heterogeneity

One of the first problems encountered by the reservoir engineer is that of organizing and utilizing the large amount of data available from core and well logging analyses. Although porosity and connate water saturation may vary aerially and vertically within a reservoir, the most important rock property variation to influence water flood performance is permeability. Permeabilities pose particular problems because they usually vary by more than an order of magnitude between different strata.

Dykstra and Parsons (1950) introduced the concept of the permeability variation V, which is designed to describe the degree of heterogeneity within the reservoir. The value of this uniformity coefficient ranges between zero for a completely homogeneous system and one for a completely heterogeneous system.

$$V = \frac{K_{50} - K_{84.1}}{K_{50}}$$

To further illustrate the use of the Dykstra and Parsons permeability variation, Craig (1971) proposed a hypothetical reservoir that consists of 10 wells (wells A through J) with detailed permeability data given for each well, as shown in Table 3-5. Each well is characterized by 10 values of permeability with each value representing 1 ft of pay.

Table 3-5 Ten-Layer Hypothetical Reservoir

Depth(ft)	A	B	C	D	E	F	G	H	I	J
6791	2.9	7.4	30.4	3.8	8.6	14.5	39.9	2.3	12.0	29.0
6792	11.3	1.7	17.6	24.6	5.5	5.3	4.8	3.0	0.6	99.0
6793	2.1	21.2	4.4	2.4	5.0	1.0	3.9	8.4	8.9	7.6
6794	167.0	1.2	2.6	22.0	11.7	6.7	74.0	25.5	1.5	5.9
6795	3.6	920.0	37.0	10.4	16.5	11.0	120.0	4.1	3.5	33.5
6796	19.5	26.6	7.8	32.0	10.7	10.0	19.0	12.4	3.3	6.5
6797	6.9	3.2	13.1	41.8	9.4	12.9	55.2	2.0	5.2	2.7
6798	50.4	35.2	0.8	18.4	20.1	27.8	22.7	47.4	4.3	66.0
6799	16.0	71.5	1.8	14.0	84.0	15.0	6.0	6.3	44.5	5.7
6800	23.5	13.5	1.5	17.0	9.8	8.1	15.4	4.6	9.1	60.0

Arranging all of these permeability values, i.e., the entire 100 permeability values, from maximum to minimum, Craig (1971) obtained the permeability distribution as shown in the log-probability scale of Figure 3-13. The resulting permeability distribution indicates that this hypothetical reservoir is characterized by a permeability variation of 70%, or:

$$V = \frac{K_{50} - K_{84.1}}{K_{50}} = \frac{10 - 3}{10} = 0.7$$

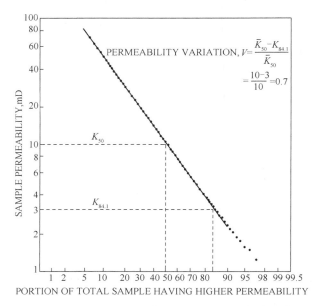

Figure 3-13 Determination of permeability variation for the hypothetical reservoir

1.2 Minimum number of layers

Based on a computer study, Craig (1971) outlined some guidelines for selecting the minimum number of layers needed to predict the performance of a reservoir under waterflooding operation. The author simulated the performance of a waterflood five-spot pattern that is composed of 100 layers with permeability variations ranging from 0.4 to 0.8. The minimum number of layers required to match results of the 100-layer model was determined as a function of mobility ratio M and permeability variation V. Tables 3-6 through 3-8 summarize results of these simulations and provide a guide to selection of the number of layers for five-spot patterns.

Table 3-6 Minimum Number of Layers for $WOR > 2.5$

Mobility Ratio	Permeability Variation							
	0.1	0.2	0.3	0.4	0.5	0.6	0.7	0.8
0.05	1	1	2	4	10	20	20	20
0.1	1	1	2	4	10	20	100	100
0.2	1	1	2	4	10	20	100	100
0.5	1	2	2	4	10	20	100	100
1.0	1	3	3	4	10	20	100	100
2.0	2	4	4	10	20	50	100	100
5.0	2	5	10	20	50	100	100	100

Table 3 – 7 Minimum Number of Layers for WOR > 5

Mobility Ratio	Permeability Variation							
	0.1	0.2	0.3	0.4	0.5	0.6	0.7	0.8
0.05	1	1	2	4	5	10	10	20
0.1	1	1	2	4	10	10	10	100
0.2	1	1	2	4	10	10	20	100
0.5	1	2	2	4	10	10	20	100
1.0	1	2	3	4	10	10	20	100
2.0	2	3	4	5	10	10	50	100
5.0	2	4	5	10	20	100	100	100

Table 3 – 8 Minimum Number of Layers for WOR > 10

Mobility Ratio	Permeability Variation							
	0.1	0.2	0.3	0.4	0.5	0.6	0.7	0.8
0.05	1	1	1	2	4	5	10	20
0.1	1	1	1	2	5	5	10	20
0.2	1	1	2	3	5	5	10	20
0.5	1	1	2	3	5	5	10	20
1.0	1	1	2	3	5	10	10	50
2.0	1	2	3	4	10	10	20	100
5.0	1	3	4	5	10	100	100	100

1.3 The zonation problem

In waterflooding calculations, it is frequently desirable to divide the reservoir into a number of layers that have equal thickness but different permeabilities and porosities. Traditionally, two methods are used in the industry to assign the proper average permeability for each layer: (1) the positional method or (2) the permeability ordering method.

1) Positional method

The positional method describes layers according to their relative location within the vertical rock column. This method assumes that the injected fluid remains in the same elevation (layer) as it moves from the injector to the producer. Miller and Lents (1966) successfully demonstrated this concept in predicting the performance of the Bodcaw Reservoir Cycling Project. The authors proposed that the average permeability in a selected layer (elevation) should be calculated by applying the geometric – average permeability:

$$K_{avg} = \exp\left(\frac{\sum_{i=1}^{n} h_i \ln K_i}{\sum_{i=1}^{n} h_i}\right)$$

If all the thicknesses are equal, then:

$$K_{avg} = (K_1 K_2 K_3 \cdots K_n)^{1/n}$$

2) Permeability ordering method

The permeability ordering method is essentially based on the Dykstra and Parsons (1950) permeability sequencing technique. The core analysis permeabilities are arranged in a decreasing permeability order and a plot like that shown in Figure 3-13 is made. The probability scale is divided into equal-percent increments with each increment representing a layer. The permeability for each layer is assigned to the permeability value that corresponds to the midpoint of each interval.

Porosity assignments for the selected reservoir layers may also be treated in a similar manner to that of the permeability ordering approach. All porosity measurements are arranged in decreasing order and a plot of the porosity versus percentage of thickness with greater porosity is made on a Cartesian-probability scale (rather than a log-probability scale). The porosity of each layer can then be obtained for each interval of thickness selected.

The permeability ordering technique is perhaps the most widely used approach in the petroleum industry when determining the vertical sweep efficiency.

2. Calculation of Vertical Sweep Efficiency

Basically two methods are traditionally used in calculating the vertical sweep efficiency E_V: (1) Stiles' method and (2) the Dykstra-Parsons method. These two methods assume that the reservoir is composed of an idealized layered system. The layered system is selected based on the permeability ordering approach with layers arranged in order of descending permeability. The common assumptions of both methods are:

(1) No cross-flow between layers;

(2) Immiscible displacement;

(3) Linear flow;

(4) The distance water has traveled through each layer is proportional to the permeability of the layer;

(5) Piston-like displacement.

The basic idea used in Stiles' method and the Dykstra-Parsons method is to determine the frontal position in each layer at the time water breakthrough occurs in successive layers. If the flow capacity of each layer is defined by the product of permeability and thickness, i.e., Kh, then the water and oil flow rates from all layers can be calculated to yield the producing water-oil ratio.

2.1 Stiles' method

Stiles (1949) proposed an approach that takes into account the effect of permeability variations in predicting the performance of waterfloods. Stiles assumes that in a layered system, the water breakthrough occurs in a sequence that starts in the layer with the highest permeability. Assuming that the reservoir is divided into n layers that are arranged in a descending permeability order with breakthrough occurring in a layer i, all layers from 1 to i have already been swept by water. The remaining layers obviously have not reached breakthrough.

Based on the above concept, Stiles proposed that the vertical sweep efficiency can be calculated from the following expression:

$$E_V = \frac{K_i \sum_{j=1}^{i} h_j + \sum_{j=i+1}^{n} (Kh)_j}{K_i h_t} \tag{3-29}$$

where i——breakthrough layer, i.e., $i = 1, 2, 3, \cdots, n$;

n——total number of layers;

E_V——vertical sweep efficiency;

h_t——total thickness, ft;

h_i——layer thickness, ft.

If the values of the porosity vary between layers, Equation can be writtenas:

$$E_V = \frac{\left(\frac{K}{\phi}\right)_i \sum_{j=1}^{i} (\phi h)_j + \sum_{j=i+1}^{n} (Kh)_j}{\left(\frac{K}{\phi}\right)_i \sum_{j=1}^{h} (\phi h)_j} \tag{3-30}$$

Stiles also developed the following expression for determining the surface water-oil ratio as breakthrough occurs in any layer:

$$WOR_s = A \frac{\sum_{j=1}^{i} (Kh)_j}{\sum_{j=i+1}^{n} (Kh)_j} \tag{3-31}$$

with

$$A = \frac{K_{rw} \mu_o B_o}{K_{ro} \mu_w B_w} \tag{3-32}$$

Both the vertical sweep efficiency and surface WOR equations are used simultaneously to describe the sequential breakthrough as it occurs in layer 1 through layer n. It is usually convenient to represent the results of these calculations graphically in terms of lg WOR_s as a function of E_V.

2.2 The Dykstra-Parsons method

Dykstra and Parsons (1950) correlated the vertical sweep efficiency with the following parameters:

(1) Permeability variation V;

(2) Mobility ratio M;

(3) Water-oil ratio WOR_r as expressed in bbl/bbl.

They presented their correlation in a graphical form for water-oil ratios of 0.1, 0.2, 0.5, 1, 2, 5, 10, 25, 50, and 100 bbl/bbl. Figure 3-14 shows Dykstra and Parsons' graphical correlation for a WOR of 50 bbl/bbl. Using a regression analysis model, De Souza and Brigham (1981) grouped the vertical sweep efficiency curves for $0 \leq M \leq 10$ and $0.3 \leq V \leq 0.8$ into one curve as shown in Figure 3-15. The authors used a combination of WOR, V, and M to define the correlation parameter Y of Figure 3-15.

$$Y = \frac{(WOR + 0.4)(18.948 - 2.499V)}{(M - 0.8094V + 1.137)10^x} \tag{3-33}$$

with

$$x = 1.6453V^2 + 0.935V - 0.6891 \tag{3-34}$$

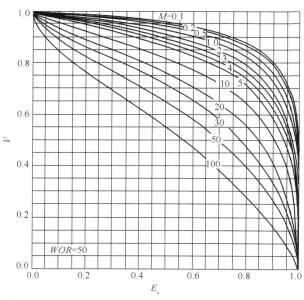

Figure 3-14 Vertical sweep efficiency curves for $WOR = 50$

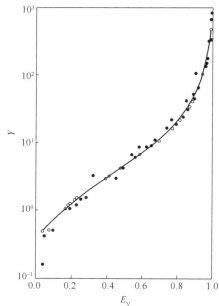

Figure 3-15 E_V versus the correlating parameter Y

The specific steps involved in determining the vertical sweep efficiency as a function of water-oil ratios are summarized below:

(1) Calculate the mobility ratio M and permeability variation V.

(2) Select several values for the WOR, e.g., 1, 2, 5, 10, and calculate the correlating parameter Y at each selected WOR.

(3) Enter Figure 3 – 15 with each value of Y and determine the corresponding values of the vertical sweep efficiency E_V.

(4) Plot WOR versus E_V.

To further simplify the calculations for determining E_V, Fassihi curve-fitted the graph of Figure 3 – 15 and proposed the following nonlinear function, which can be solved iteratively for the vertical sweep efficiency E_V:

$$a_1 E_V^{a_2} (1 - E_V)^{a_3} - Y = 0 \qquad (3-35)$$

where, $a_1 = 3.334088568, a_2 = 0.7737348199, a_3 = 1.225859406$.

The Newton-Raphson method is perhaps the appropriate technique for solving Equation (3 – 35). To avoid the iterative process, the following expression could be used to estimate the vertical sweep efficiency using the correlating parameter Y.

$$E_V = a_1 + a_2 \ln Y + a_3 (\ln Y)^2 + a_4 (\ln Y)^3 + a_5 / \ln Y + a_6 Y$$

with the coefficients a_1 through a_6 as given by:

$a_1 = 0.19862608$ $\qquad a_2 = 0.18147754$

$a_3 = 0.01609715$ $\qquad a_4 = -4.6226385 \times 10^{-3}$

$a_5 = -4.2968246 \times 10^{-4}$ $\qquad a_6 = 2.7688363 \times 10^{-4}$

Section 3 Method to Predict Recovery Performance for Layered Reservoir

To account for the reservoir vertical heterogeneity when predicting reservoir performance, the reservoir is represented by a series of layers with no vertical communication, i.e., no crossflow between layers. Each layer is characterized by a thickness h, permeability K, and porosity ϕ. The heterogeneity of the entire reservoir is usually described by the permeability variation parameter V. Three of the methods that are designed to predict the performance of layered reservoirs are discussed below.

1. Simplified Dykstra-Parsons Method

Dykstra and Parsons (1950) proposed a correlation for predicting waterflood oil recovery that uses the mobility ratio, permeability variation, and producing water-oil ratio as correlating parameters. Johnson (1956) developed a simplified graphical approach for the Dykstra and Parsons method that is based on predicting the overall oil recovery R at water-oil ratios of 1, 5, 25, and 100 bbl/bbl. Figure 3-16 shows the proposed graphical charts for the four selected WOR_s.

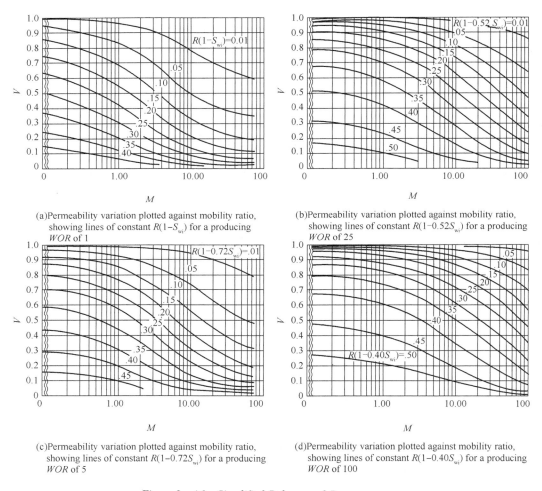

Figure 3-16 Simplified Dykstra and Parsons curves

The correlating parameters shown in Figure 3-16, are: R = overall oil recovery factor, S_{wi} = initial water saturation, M = mobility ratio, V = permeability variation.

The practical application of the simplified Dykstra and Parsons method is outlined below:

(1) Calculate the permeability variation V and mobility ratio M.

(2) Using the permeability ratio and mobility ratio, calculate the overall oil recovery factor R from the four charts at WOR_s of 1, 5, 25, 100 bbl/bbl. For example, to determine the oil recovery factor when the WOR_s reaches 5 bbl/bbl for a flood pattern that is characterized by a V and M of 0.5 and 2, respectively:

①Enter the appropriate graph with these values, i.e., 0.5 and 2.

②The point of intersection shows that $R(1 - 0.72 S_{wi}) = 0.25$.

③If the initial water saturation S_{wi} is 0.21, solve for the recovery factor to give $R = 0.29$.

(3) Calculate the cumulative oil production N_p at each of the four water-oil ratios, i.e., 1, 5, 25, and 100 bbl/bbl.

(4) If the water-oil ratio is plotted against the oil recovery on semi-log paper and a Cartesian scale, the oil recovery at breakthrough can be found by extrapolating the line to a very low value of WOR_s, as shown schematically in Figure 3-17.

(5) For a constant injection rate, adding the fill-up volume W_{if} to the cumulative oil produced at breakthrough and dividing by the injection rate can estimate the time to breakthrough.

(6) The cumulative water produced at any given value of WOR_s is obtained by finding the area under the curve of WOR_s versus N_p, as shown schematically in Figure 3-18.

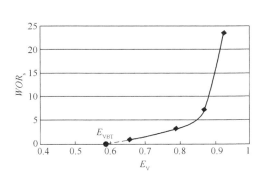
Figure 3-17 WOR_s versus E_V relationship

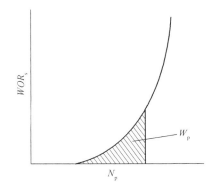
Figure 3-18 Cumulative water production from $WOR_s - N_p$ curve

(7) The cumulative water injected at any given value of WOR_s is calculated by adding cumulative oil produced to the produced water and fill-up volume, or:

$$W_{inj} = N_p B_o + W_p B_w + W_{if}$$

2. Modified Dykstra-Parsons Method

Felsenthal, Cobb, and Heuer (1962) extend the work of Dykstra and Parsons to account for the presence of initial gas saturation at the start of flood. Assuming a constant water injection rate i_w, the method is summarized in the following steps:

Step 1. Perform the following preliminary calculations to determine:

(1) Pore volume PV and oil in place at start of flood N_S;

(2) Water cut f_w as a function of S_w;

(3) Slope (df_w/dS_w) as a function of S_w;

(4) Average water saturation at breakthrough $\bar{S}_{wBT}$;

(5) Mobility ratio M;

(6) Displacement efficiency at breakthrough E_{DBT};

(7) Areal sweep efficiency at breakthrough E_{ABT};

(8) Permeability variation V;

(9) Fill-up volume W_{if}.

Step 2. Calculate the vertical sweep efficiency at assumed water-oil ratios of 1, 2, 5, 10, 15, 20, 25, 50, and 100 bbl/bbl.

Step 3. Plot WOR versus E_V on a Cartesian scale, and determine the vertical sweep efficiency at breakthrough E_{VBT} by extrapolating the WOR versus E_V curve to $WOR = 0$.

Step 4. Calculate cumulative water injected at breakthrough:

$$W_{iBT} = PV(\bar{S}_{wBT} - S_{wi})E_{ABT}E_{VBT}$$

Step 5. Calculate cumulative oil produced at breakthrough from the following expression:

$$(N_p)_{BT} = \frac{W_{iBT} - W_{if}E_{VBT}}{B_o}$$

Step 6. Calculate the time to breakthrough t_{BT} from:

$$t_{BT} = \frac{W_{iBT}}{i_W}$$

Step 7. Assume several values for water-oil ratios WOR_r, e.g., 1, 2, 5, 10, 15, 20, 25, 50, and 100 bbl/bbl.

Step 8. Determine E_V for each assumed value of WOR_r (see step 3).

Step 9. Convert the assumed values of WOR_r to water cut f_{w2} and surface WOR, respectively:

$$f_{w2} = \frac{WOR_r}{WOR_r + 1}$$

$$WOR_s = WOR_r \left(\frac{B_o}{B_w}\right)$$

Step 10. Determine the water saturation S_{w2} for each value of f_{w2} from the water cut curve.

Chapter 3　Method to Calculate Sweep Efficiency

Step 11. Determine the areal sweep efficiency E_A for each value of f_{w2}.

Step 12. Determine the vertical sweep efficiency E_V for each value of f_{w2}.

Step 13. Determine the average water saturation $\overline{S}_{w2}$ for each value of f_{w2}.

Step 14. Calculate the displacement efficiency E_D for each $\overline{S}_{w2}$ in step 13.

Step 15. Calculate cumulative oil production for each WOR_s from:

$$N_p = N_S E_D E_A E_V - \frac{PVS_{gi}(1 - E_A E_V)}{B_o}$$

Step 16. Plot the cumulative oil production N_p versus WOR_s on Cartesian coordinate paper, and calculate the area under the curve at several values of WOR_s. The area under the curve represents the cumulative water production W_p at any specified WOR_s, i.e., $(W_p)_{WOR}$.

Step 17. Calculate the cumulative water injected W_{inj} at each selected WOR from:

$$W_{inj} = (N_p)_{WOR} B_o + PVS_{gi}(E_V)_{WOR}$$

where　$(N_p)_{WOR}$——cumulative oil production when the water-oil ratio reaches WOR, STB;

$(E_V)_{WOR}$——vertical sweep efficiency when the water-oil ratio reaches WOR.

Step 18. Calculate the time to inject W_{inj}:

$$t = \frac{W_{inj}}{i_w}$$

Step 19. Calculate the oil and water flow rates, respectively:

$$Q_o = \frac{i_w}{B_o + B_w WOR_s}$$

$$Q_w = Q_o WOR_s$$

3. Craig-Geffen-Morse Method

With the obvious difficulty of incorporating the vertical sweep efficiency in oil recovery calculations, Craig et al. (1955) proposed performing the calculations for only one selected layer in the multilayered system. The selected layer, identified as the base layer, is considered to have a 100% vertical sweep efficiency. The performance of each of the remaining layers can be obtained by "sliding the timescale" as summarized in the following steps:

Step 1. Divide the reservoir into the appropriate number of layers.

Step 2. Calculate the performance of a single layer, i.e., the base layer, for example, layer n.

Step 3. Plot cumulative liquid volumes (N_p, W_p, W_{inj}) and liquid rates (Q_o, Q_w, i_w) as a function of time t for the base layer, i.e., layer n.

Step 4. For each layer (including the base layer n) obtain: K/ϕ、ϕh、Kh.

Step 5. To obtain the performance of layer i, select a succession of times t and obtain plotted values N_p^*、W_p^*、W_{inj}^*、Q_o^*、Q_w^* 和 i_w^* by reading the graph of step 3 at time t^*:

$$t_i^* = t \frac{\left(\frac{K}{\phi}\right)_i}{\left(\frac{K}{\phi}\right)_n}$$

Then calculate the performance of layer i at any time t from:

$$N_p = N_p^* \frac{(\phi h)_i}{(\phi h)_n}$$

$$W_p = W_p^* \frac{(\phi h)_i}{(\phi h)_n}$$

$$W_{inj} = W_{inj}^* \frac{(\phi h)_i}{(\phi h)_n}$$

$$Q_o = Q_o^* \frac{(K/\phi)_i}{(K/\phi)_n}$$

$$Q_w = Q_w^* \frac{(K/\phi)_i}{(K/\phi)_n}$$

$$i_w = i_w^* \frac{(K/\phi)_i}{(K/\phi)_n}$$

Step 6. The composite performance of the flood pattern at time t is obtained by summation of individual layer values.

Chapter 4 Empirical Methods for Dynamic Analysis

General working procedure for empirical methods is: Systematically observe the production dynamics of the oilfield, accurately and comprehensively collect the data that can explain the production rules, deeply analyze the data, and discover the things with regularity, then the data with regularity is processed mathematically. Give empirical equations to express these laws (including the determination of empirical parameters), and finally use the empirical rules that have been summarized to predict future production dynamics.

Section 1 Oil Production Decline Law

Regardless of the types of reservoir or the driving types of oil and gas field, as the development progresses, it will enter the stage of production decline. According to the data of production and cumulative production at this stage, the method of production decline analysis can not only predict the changes of oilfield production and cumulative production in the future, but also can effectively predict the recoverable reserves and remaining recoverable reserves of the oilfield. This section will introduce the analysis method of oilfield production decline law.

1. Production Change Pattern

The production change of oil and gas fields (or oil and gas reservoirs) is generally divided into three stages: the rising production period, the stable production period and the decline period, as shown in Figure 4 – 1.

1.1 Rising production period

At the beginning of the development of oil and gas fields, new wells have been put into production continuously, and production facilities have been continuously improved. Therefore, the

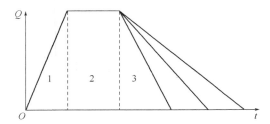

Figure 4-1 Oil and gas field production change pattern

1—Rising production period; 2—Stable production period; 3—Decline period

output has also increased year by year. This stage is called the rising period of oil and gas fields, referred to as the rising production period. The rising period is usually short, ranging from 2 to 5 years. The specific time is related to the oilfield size, geological, political, economic and technical conditions. Since only 5% ~ 10% of the recoverable geological reserves can be produced at this stage, and the time is short, which is mainly interfered by human factors, the law of production change in the rising production period is rarely studied.

1.2 Stable production period

After the end of the rising production period, the constructions of well network and pipe network system for oilfield development have been basically completed, and the overall production facilities for oilfield development have been basically established. The systems of oil-gas-water separation and gathering have started to operate normally, and the injection system has begun to work normally. In the next period of time, oil and gas production will enter full-load operation period and will reach the maximum design capacity. This stage is called the stable production period of oil and gas field development. The stable production period is a golden period for oil and gas field development. The length of this stage is mainly affected by the geological conditions of the reservoir and the development system. Under normal circumstances, the larger the reserve size it is, the longer the stable production period. According to oil industry standards, for small and medium sized oil and gas fields, the stable production period should be about 2 to 5 years; for large and medium-sized oil and gas fields, it should be about 5 to 10 years; for large-scale oil and gas fields, it is generally more than 10 years. For example, Daqing Oilfield has stable annual oil production of 50 million tons for 27 years. How to determine whether the stable production period of an oil and gas field is appropriate? It should be designed through the principle of maximizing economic benefits. The development benefit of the stable production period is the best, and generally about 50% of the recoverable geological reserves can be produced. Its length is represented by T and its calculation equation is as follows:

$$T = \frac{NR_o}{Q}$$

where N——Geological reserves;

        R_o——Recovery degree during the stable production period;

        Q——Annual crude oil production during the stable production period.

It can be seen from the above equation that the higher the output of the oil and gas field during the stable production period, the shorter the stable production period.

1.3 Decline period

After the end of a stable production period, it comes into the decline period. The decline period is the stage of exploitation that no oil and gas field can avoid. The emergence of the declining period is a sign that the effective driving energy of the formation is tending to be exhausted. For the natural energy driven reservoirs, during the stable production period, the formation energy has been exhausted. Only the production can be reduced to achieve a new balance of the low driving energy level. Water injection development or natural water driven reservoirs have been stabilized and most of the oil wells are water breakthrough. After water production, the water drive efficiency will decrease and the oil and gas production will decline. Unlike the periods of the rising and stable production, the production decline is a natural process in this period. However, due to the differences in geological conditions between oil and gas fields and the different development system settings, the production decline path and the decrement mode are different (Figure 4 – 1). This requires a certain amount of research work to find out the specific decline laws, the main causes and contradictions affecting oil and gas production for specific oil fields, so as to predict and plan oil and gas production and improve the economic benefits of oil and gas field development.

The duration of the decline period is mainly affected by the geological conditions of the oil and gas fields and the economic and technological conditions at that time. Most oil and gas fields have a long period of decline in production, generally 10 to 30 years. In this period, up to 40% to 50% of the recoverable geological reserves can be extracted.

As the oil and gas production continues to decrease in the decline period, the economic benefits of the oilfield continue to decrease. In order to improve economic efficiency, the following measures will be taken to slow down the production decline: subdivided development layers, adjustment of well network, adjustment of water injection and production system, EOR methods, measures to increase production and injection. After all the measures, the oil and gas production still can not bring about economic benefits, the oil and gas production process will be terminated, and the oil and gas fields will be abandoned eventually.

2. Law of Oil Production Decline

The methods for studying the law of production decline are generally: First, plot the relationship between production and time, or the relationship between production and cumulative production. Then choose a coordinate to turn the curve of the decreasing part of production into a straight line (or close to a straight line), and write down the equation of the straight line, which can be used to find the empirical relationship between production and time, and then predict the future dynamic indicators of the oil field. The law of oil production decline is common in the oil field, especially the exponential decline law and hyperbolic decline law.

2.1 Production decline rate

In the process of oilfield development, with the change of underground energy and the decrease of recoverable reserves, the oil production always decreases, and the general decline rate indicates the decline rate of production. The so-called decline rate refers to the change in unit output per unit time, usually expressed as a decimal or a percentage. According to the statistical analysis of the actual data of the oilfield, the decline rate can be expressed as the following form.

$$D = -\frac{dq}{qdt} = kq^n \qquad (4-1)$$

where D——Production decline rate;

q——Production;

t——Time;

n——Declining exponent ($0 \leqslant n \leqslant 1$);

k——Proportional constant.

The negative sign in equation (4-1) indicates that the output is declining as the development time increases.

[Example 4-1] According to the decline rate equation, calculate the natural decline rate and comprehensive decline rate of the well group in 2020 using the data given in Table 4-1.

Table 4-1 Output data of the well group

Time, year	Annual liquid production, t	Annual oil production, t	Annual oil production increment of measures, t
2019	83976	7222	126
2020	100450	6630	94

Solution: When we calculate the natural decline rate of this well group in 2020, it is necessary to deduct 94t of oil increment of measures from annual oil production. That is,

Natural decline rate = $[7222 - (6630 - 94)] \div 7222 = 9.5\% \, a^{-1}$

Comprehensive decline rate = $(7222 - 6630) \div 7222 = 8.2\% \, a^{-1}$

So the natural decline rate in 2020 is $9.5\% \, a^{-1}$ and the comprehensive decline rate is $8.2\% \, a^{-1}$.

Conclusion: In fact, it is commonly used to depict the speed of production decline in the oilfield. Generally, when $D < 0.1 a^{-1}$, the production decreases slowly; when $D = 0.1 a^{-1} \sim 0.3 a^{-1}$, the production decline is medium; when $D > 0.3 a^{-1}$, the production decreases rapidly.

The rate of production decline is influenced by many factors, and the factors vary from reservoirs to reservoirs. Normally, the main objective factors of affecting production decline involve the controllable reserves per well, natural energy supply rate and water cut increasing rate. The main subjective factors include oil recovery speed and artificial energy supply speed.

2.2 The related formulas of production decline law

There are three types of production decline law: hyperbolic, exponential and harmonic, and the related formulas are shown in Table 4-2.

Table 4-2 The related formulas of production decline law

decline type	basic characteristics	Basic relation formula $q_t - t$	$N_p - t$	$N_p - q_t$	Maximum cumulative production
exponential	$n = 0$, $D = D_i$	$q_t = q_i \exp(-D_i t)$	$N_p = \dfrac{q_i}{D_i}[1 - \exp(-D_i t)]$	$N_p = \dfrac{1}{D_i}(q_i - q_t)$	$N_p = \dfrac{q_i}{D_i}$
hyperbolic	$0 < n < 1$, $D < D_i$	$q_t = q_i(1 + nD_i t)^{-\frac{1}{n}}$	$N_p = \dfrac{q_i}{(n-1)D_i}[(1+nD_i t)^{\frac{n-1}{n}} - 1]$	$N_p = \dfrac{q_i^n}{(1-n)D_i}[q_i^{1-n} - q_t^{1-n}]$	$N_p = \dfrac{q_i}{(1-n)D_i}$
harmonic	$n = 1$, $D < D_i$	$q_t = q_i(1 + D_i t)^{-1}$	$N_p = \dfrac{q_i}{D_i}\ln(1 + D_i t)$	$N_p = \dfrac{q_i}{D_i}\ln\dfrac{q_i}{q_t}$	$N_p = \dfrac{q_i}{D_i}\ln q_i$

2.3 Comparison of three types of production decline law

1) When $n = 0$, $D_t = D_i = $ constant. That is to say, the decline rate of exponential type is constant, also called constant percentage decrement.

2) When $n = 1$, $D_t = D_i \dfrac{q_t}{q_i}$, since $\dfrac{q_t}{q_i} < 1$, the decline rate of harmonic type decreases with the decrease of the production, that is, as the development time goes by, the decline rate gradually slows down.

3) When $0 < n < 1$, $D_t = D_i \left(\dfrac{q_t}{q_i}\right)^n$. Since $\left(\dfrac{q_t}{q_i}\right)^n < 1$, the decline rate of hyperbolic type decreases as the production decreases.

According to the above evidence, under the same initial decline rate, the decline speed of exponential type is the fastest.

Since $\dfrac{q_t}{q_i} < 1$ and $0 < n < 1$, it is known $\left(\dfrac{q_t}{q_i}\right)^n > \dfrac{q_t}{q_i}$, so the decline rate of hyperbolic type is larger than that of harmonic type.

Above all, the speed of decline in production is mainly determined by the declining exponent n and the initial decline rate D_i. When the initial decline rate and the initial output of the declining period are the same, the decline speed of the exponential type decline is the fastest, that of the hyperbolic type decline is medium, and that of the harmonic type decline is the slowest. When the decline type is determined, the larger the initial decline rate, the faster the yield decline. A typical curve for the three types of decrement is shown in Figure 4-2.

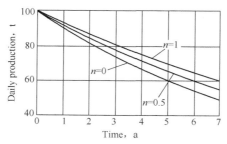

Figure 4-2 Typical decline curve

2.4 Discrimination of production decline type

Based on the previous theoretical analysis, here we mainly introduce four discriminant methods: graphic method, trial and error method (trial-and-error method), typical curve matching method and diagnostic curve method.

1) Graphic method

(1) Take $\lg q_t$ as the ordinate and t as the abscissa to plot the curve, as shown in Figure 4-3. It is said to be followed exponential decline if it is a straight line.

(2) Take $\lg q_t$ as the ordinate and N_p as the abscissa to plot the curve, as shown in Figure 4-4. It is said to be subject to harmonic decline if it is a straight line. Otherwise, the hyperbolic decline is obeyed.

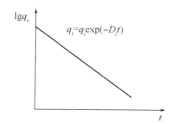 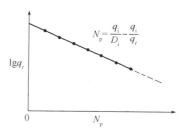

Figure 4-3 Exponential type decline law   Figure 4-4 Harmonic type decline law

2) Trial and error method

When the production decline law satisfies neither exponential type nor harmonic type, it can only be hyperbolic type. But their parameters cannot be determined by linear regression analysis, but can only be solved and judged by trial and error method.

Deformed by the equation of hyperbolic type in Table 4 – 2:

$$\left(\frac{q_i}{q_t}\right)^n = a + bt \qquad (4-2)$$

where $\qquad a = 1, b = nD_i$

Assign a value to the declining exponent n at first. If the data regression satisfies a good linear relationship, the value of n can be viewed as accurate. The value of n is larger if the curve is upwarped, on the contrary, the value of n is smaller when the curve is downward (that is, horizontal), as shown in Figure 4 – 5. Adjust the value of n until the data meets the linear relationship, so as to determine the value of n and the values of a and b, then determine the value of D_i.

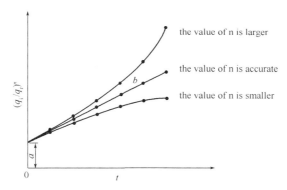

Figure 4 – 5   Trial calculation diagram of hyperbolic decline law

3) Typical curve matching method

Deformation according to equations in Table 4 – 2,

$$\frac{q_i}{q_t} = \exp(D_i t) \quad (n = 0, \text{Exponential type decline law})$$

$$\frac{q_i}{q_t} = (1 + nD_i t)^{\frac{1}{n}} \quad (0 < n < 1, \text{Hyperbolic type decline law})$$

$$\frac{q_i}{q_t} = 1 + D_i t \quad (n = 1, \text{Harmonic type decline law})$$

The specific steps of this method are as follows:

(1) Draw the theoretical chart (give different values of the declining exponent n and $D_i t$, and make curves between q_i/q_t and $D_i t$), as shown in Figure 4 – 6;

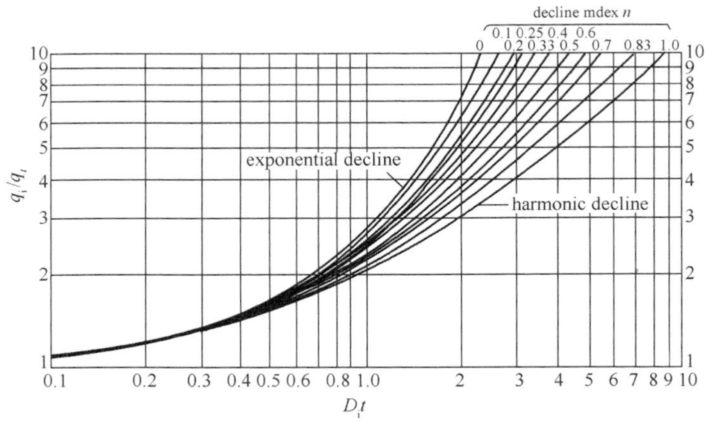

Figure 4-6　Diagram of typical curve fitting theory

(2) Draw the actual curve (the relation curves between q_i/q_t and t in the decline stage with transparent paper);

(3) By the contrast between the actual curve with the theoretical chart, the value of n can be read directly, and the value of D_i can be calculated from a data point where the actual curve overlaps with the theoretical chart. The calculation formula is as follows:

$$D_i = \frac{D_i t(\text{theoretical})}{t(\text{practical})} \tag{4-3}$$

4) Diagnostic curve of production decline law

The decline rate of production is generally not a constant, but a variable varying with production. On the basis of massive statistical laws, the decline rate meets the following requirements:

$$D = Kq^n \tag{4-4}$$

Take logarithms on both sides to obtain:

$$\lg D = \lg K + n\lg q \tag{4-5}$$

According to the definition of decline rate, the following formula is obtained:

$$D = -\frac{dq}{qdt}$$

Then draw the relevant curve between decline rate and output in the double logarithmic coordinate system according to equation (4-5) after calculating the decline rate corresponding to each output, as shown in Figure 4-7. It is clear that the logarithmic curve between the decline rate and output is a straight line in the figure at a certain stage of output change. And the slope of the straight line is the decline index n. The production points of the same straight line meet the same decline type. Li Chuanliang called this curve the diagnostic curve of the production decline law and equation (4 – 5) the diagnostic equation of production decline law.

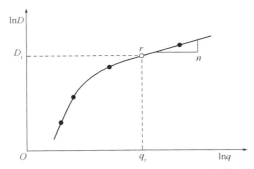

Figure 4 – 7   Diagnostic curve of production decline rule

Section 2   Rising Law of Water Cut

For water flooding oilfields, whether it relies on artifical water injection or natural water flooding, after the end of the waterless oil recovery period, oil production containing water will be carried out for a long time, and the water cut will gradually increase, which is an important factor affecting the stable production of oilfields. The water cut of water flooding oilfield is a comprehensive index influenced by many factors in oilfield development. It not only reflects the restriction of oil layer and crude oil physical property on the oil-water movement law in oil reservoir, but also reflects the effects of various technical measures in the mining process. Therefore, for such oilfields, the actual production data in oilfield exploitation is used to analyze the rising law of water cut, to study the geological engineering factors affecting the increase of water cut, and to formulate practical measures for controlling water and stabilizing oil and controlling water and oil in different production stages. It is a regular and extremely important task in the development of water flooding oilfields. The article will mainly introduce the water rising law and application of water flooding oilfields.

1. Change Law of Water Cut

In general, a basic curve representing the development of a water flooding oilfield is the relationship between water cut and recovery, as shown in Figure 4 – 8. The shape and location of this curve comprehensively reflect the geological characteristics of the reservoir, the distribution of oil and water, and the nature, development methods and process measures. From the perspective of research work, this curve is a curve with a special shape, which is difficult to express with a simple formula. Therefore, when studying the law of water cut, it is necessary to perform certain mathematical processing and transformation on these empirical data.

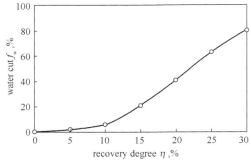

Figure 4-8 Curve of water cut and recovery degree of an oil field

Production practice shows that a natural water flooding or artificial water flooding reservoir, when it has been fully developed and entered stable production, when its water cut reaches a certain level and gradually rises, on a single logarithmic coordinate paper, to accumulate water production, the logarithm is the ordinate, and the cumulative oil production (or the degree of recovery) is the abscissa, then the relationship between the two is a straight line, which is called the water flooding characteristic curve.

Figure 4-9 shows the water drive curve of water injection development in an oilfield in China. This line generally begins from the mid-water phase (water cut is around 20%) and remains unchanged during the high water cut period. When the mining method of injection-production well pattern, injection-production layer system and injection-production intensity remains unchanged, the linear property remains unchanged. When injection production system is changed, an inflection point appears, but the linear relationship still holds. As shown in Figure 4-9, when the water cut reaches about 47.7%, there is a turning point in the straight line, which is due to the certain adjustment measures at this time.

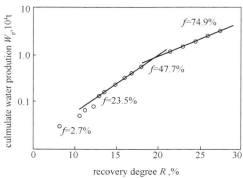

Figure 4-9 Changes of water flooding characteristic curve before and after an oilfield adjustment measure

It is very important to discover this law, because with such a law, the water content law of the oilfield can be correctly expressed. This single logarithmic relationship between oil and water production can be seen in many oil fields at home and abroad, and has a fairly widespread universality. Most of the water injection development oil fields in China also conform to this law. In this way, people can use this quantitative law to describe and predict the water cut, oil production, ultimate recovery and recoverable reserves of various oil fields in the production process.

The current consensus is that there are significant differences in the characteristics of water flooding with different oil-water viscosities. For low-viscosity oilfields, the oil-water viscosity ratio is low, and the rate of water cut rises slowly in the initial stage of development, in terms of water cut and recovery. The relationship curve is concave, the main reserves are produced in the middle and low water cut period; while the medium and high viscosity oil field is opposite, the convex curve is shown on the relationship between the water cut and the recovery degree, and the main reserves are produced in the high water cut period. This is due to the non-piston nature of the water flooding. The wettability and heterogeneity of the reservoirs exacerbate this difference.

The current general rate of water content classification criteria are as follows.

(1) Water-free oil recovery period: water cut is less than 2%;

(2) Low water cut oil recovery period: water cut is 2% to 20%;

(3) Middle water cut oil recovery period: water cut is 20% ~ 60%;

(4) High water cut oil recovery period: water cut is 60% ~ 90%;

(5) Extra high water cut oil recovery period: water cut is greater than 90%;

(6) Limit water cut oil recovery period: water cut is equal to 98%.

2. Common Water Flooding Characteristics Curve

The research and application of the water-drive law curve have gone through three stages, and each stage has its own typical research methods.

2.1 Relationship between water cut and time or water cut and degree of recovery

In 1959, when the former Soviet scholar Maximov determined the recoverable reserves at the end of the water-flooding reservoir, he used some old oilfield data in the Grozne oil area to study, and considered that in the end of the water-flooding field, for a layer system, there is a statistical relationship between cumulative oil production N_p and cumulative water production W_p. The formula is

$$W_p = a_1 \exp(b_1 N_p) \tag{4-6}$$

where a_1, b_1 ——the undetermined coefficient related to the geological and development factors of the oilfield.

In 1971, American scholar Timmanman calculated the actual data of some water-flooding oilfields in the United States, pointing out that the oil-water ratio F_{ow} and cumulative oil production N_p are in a straight line relationship in the semi-logarithmic coordinate system. The formula is

$$\lg F_{ow} = b_2 N_p + a_2 \qquad (4-7)$$

within

$$F_{ow} = q_o / q_w$$

where F_{ow}——oil-water ratio;

a_2, b_2——constant.

Their research pointed out that there is a certain quantitative relationship between oil and water production changes, but its application is not sufficiently researched. Therefore, at this stage, the old method is still used on the mine to determine the relationship between water cut f_w and time t, or water cut f_w and degree of recovery R. This method is simple and intuitive. Until today, the mine is still in use, but due to many influencing factors, the actual data tends to fluctuate greatly and the regularity is not strong.

2.2 Water-drive law curve

Since the 1970s, as oilfields have generally developed water flooding, many researchers engaged in reservoir dynamics analysis have discovered that a natural water drive reservoir or an artificial water flooding reservoir has a certain water cut when it has been fully developed and entered a stable production stage, with the rising of water cut, on the semi-logarithmic coordinate paper, the cumulative water production W_p is y-coordinate (by logarithmic coordinates), and the cumulative oil production N_p is x-coordinate (by ordinary coordinates), their relationship curve is often an approximate line. The straight line segment is shown in Figure 4 – 10. This kind of curve is called the water-drive law curve (later called Type A water drive curve), which has the following two basic expressions:

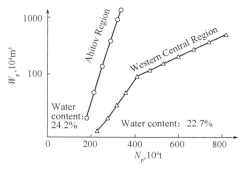

Figure 4 – 10 Curve of cumulative water production and cumulative oil production

$$\lg W_p = BN_p + A \qquad (4-8)$$

$$R = B\lg\left(\frac{f_w}{1-f_w}\right) + A \qquad (4-9)$$

where R——degree of recovery;

   f_w——water cut.

In practice, people call this method the displacement feature method. Further subdivision, equation (4-8) is called relationship curve method of the cumulative water production and the cumulative oil production; equation (4-9) is called relationship curve method of the recovery and water-oil ratio. The two types are essentially the same and can be deduced from each other.

The water-drive curve has the following two characteristics:

(1) The general oil field begins to appear straight line before and after 20% water cut.

(2) When major measures (such as fracturing) or changes in development conditions (such as layer adjustment) are taken on the oil layer, the development effect is abrupt, and the straight line segment turns.

At present, people use the water-drive law curve on the mine to judge or compare the oilfield (or oil well) mining effect, to judge the oil well water level and the incoming water direction or effective direction, and to predict the reservoir dynamic index.

2.3 Drive series formula

Based on the flooding development method commonly used in China's oilfields, the flooding characteristics of different oil reservoirs were investigated. The displacement characteristics method is applicable to oilfields with medium rock and fluid properties, and the water cut varies according to the degree of recovery. Oilfields with good physical properties or poor fluid properties are not suitable or very suitable. Practice has proved that oil fields of various properties have different types of displacement characteristics. China Petroleum Engineer Wan Jiye discusses the relationship between the water cut-recovery relationship curve and the rock pore structure, fluid properties and wettability of the oil reservoir, and can be divided into five types of displacement characteristics according to their characteristics (Figure 4-11). The series, expressed as:

(1) Convex curve:

$$R = A + B\ln(1 - f_w) \qquad (4-10)$$

(2) Convex and S inter-model transition curve:

$$\ln(1 - R) = A + B\lg(1 - f_w) \qquad (4-11)$$

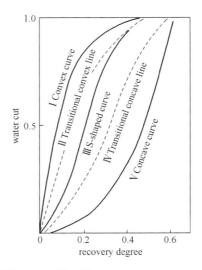

Figure 4 – 11   Various water cut curves

(3) The curve of S type:

$$R = A + B\ln\frac{f_w}{1-f_w} \tag{4-12}$$

(4) Transition curve between S type and concave type:

$$\ln R = A + Bf_w \tag{4-13}$$

(5) Concave curve:

$$\ln R = A + B\ln f_w \tag{4-14}$$

2.4 Other statistical relations

There are a lot of mathematical models to study oilfield water drive curve. The following are some common equations of statistical curves.

(1) Equation 1 (Type B water drive curve):

$$\ln L_p = a + bN_p \tag{4-15}$$

within
$$L_p = W_p + N_p$$

where L_p——cumulative fluid production.

(2) Equation 2 (Type C water drive curve):

$$\frac{L_p}{N_p} = a + bL_p \tag{4-16}$$

(3) Equation 3 (Type D water drive curve):

$$\frac{L_p}{N_p} = a + bW_p \tag{4-17}$$

(4) Equation 4:
$$\ln R_{wo} = a + bN_p \quad (4-18)$$

(5) Equation 5:
$$\ln R_{Lo} = a + bN_p \quad (4-19)$$

where R_{Lo}——production fluid-oilratio, which is the ratio of fluid production to oil production.

(6) Equation 6:
$$\ln f_w = a + bN_p \quad (4-20)$$

(7) Equation 7:
$$\ln f_w = a + b\ln N_p \quad (4-21)$$

3. The Application of Type A Water Drive Curve

3.1 Theoretical derivation

Type A water drive curve always is deformed in practical application. Equation (4 – 8) can be transformed into
$$N_p = a(\lg W_p - \lg b) \quad (4-22)$$

As mentioned above, after the development of the water flooding oilfield to a certain stage, in the semi-logarithmic coordinate system, the cumulative oil production is the common coordinate, and the cumulative water cut is the logarithmic coordinate, and the water-drive law curve is drawn as a straight line, as shown in Figure 4 – 12, the slope can be obtained by the following equation:
$$\frac{1}{a} = \frac{\lg W_{p2} - \lg W_{p1}}{N_{p2} - N_{p1}} \quad (4-23)$$

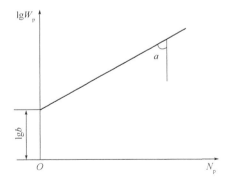

Figure 4 – 12 Typical water-drive curve

Extend the straight line segment to intersect the vertical axis to obtain the intercept (b). Finding a constant (a,b). You can develop a dynamic prediction.

1) Prediction of water production

Transform equation (4-22) into

$$W_p = b \cdot 10^{(N_p/a)} \quad (4-24)$$

or

$$W_p = 10^{(\lg b + N_p/a)} \quad (4-25)$$

According to equation (4-24) or equation (4-25), the water production can be obtained under the given production (N_p) conditions.

2) Prediction of water cut

Take the differential of equation (4-22):

$$\frac{dN_p}{dt} = a \lg e \frac{1}{W_p} \frac{dW_p}{dt}$$

By $\frac{dN_p}{dt} = q_o$, $\frac{dW_p}{dt} = q_w$, substituting the above equation to get the water to oil ratio (F_{wo}):

$$F_{wo} = \frac{q_w}{q_o} = \frac{2.3 W_p}{a} \quad (4-26)$$

The water cut can be obtained according to the following equation:

$$f_w = \frac{q_w}{q_w + q_o} = \frac{1}{1 + q_o/q_w}$$

Substituting equation (4-26) into the above equation

$$f_w = \frac{2.3 W_p}{a + 2.3 W_p} \quad (4-27)$$

According to equation (4-27), the water cut can be predicted.

3) Predicted ultimate recovery

At present, the concept of water cut limit or limit water-oil ratio is widely used in water flooding oilfields. Exceeding this limit, the oilfield loses its actual mining value. The degree of recovery achieved by reaching this limit is the ultimate recovery of the field. A typical universal water limit is 98% or a limit water-oil ratio of 49. The recovery factor values predicted by empirical methods are generally more in line with production practices than other methods.

Deformation of equation (4-26)

$$W_p = \frac{a F_{wo}}{2.3}$$

Substitute equation (4-22)

$$N_p = a\left(\lg \frac{a F_{wo}}{2.3} - \lg b\right) \quad (4-28)$$

Put the extreme water-oil ratio = 49 into the equation (4 − 28)

$$N_{pmax} = a(\lg 21.3a - \lg b) \quad (4-29)$$

When analyzing the effects of oilfield development in production, it is often required to know the relationship between water cut and the degree of recovery in order to facilitate comparison. To this end, the following three dimensionless quantities are obtained:

$$R = N_p/N$$
$$a_D = a/N$$
$$b_D = b/N$$

In the equation N is the geological reserve, then there is

$$R_{max} = \frac{N_{pmax}}{N} = \frac{a}{N}(\lg 21.3a - \lg b) \quad (4-30)$$

or

$$R_{max} = a_D(\lg 21.3a_D - \lg b_D) \quad (4-31)$$

The maximum recovery rate R_{max} obtained according to the above equation is the ultimate recovery of the oil field.

3.2 Practical application

[**Example 4 − 2**] It is known that a certain oilfield oil layer is developed with water injection, and its production data is shown in Table 4 − 3. The oilfield geological reserves are 2409×10^4 t. The water cut corresponding to the cumulative oil recovery is calculated according to the water-drive law, and the final oil recovery and recovery factor when the water cut reaches 98% is obtained.

Table 4 − 3 Oilfield production data and calculation results

Serial number	Time	Production Data			calculate data		
		Cumulative oil production 10^4 t	cumulative water production $10^4 m^3$	water cut %	cumulative water production $10^4 m^3$	water cut %	degree of recovery %
1	1959	388.7	32.9	20.8	42.7	29	16.09
2	1960	434.4	58.2	35.1	66.0	38.8	18.03
3	1961	465.7	82.0	43.3	89.0	46	19.33
4	1962	490.2	105.7	49.0	112	49	20.43
5	1963	508.9	125.6	51.3	135	56	21.12
6	1964	526.9	151.0	58.6	159	60	21.87
7	1965	544.1	180.9	63.4	186	65	22.59
8	1966	562.4	214.5	64.9	224	68	23.34

continued

Serial number	Time	Production Data			calculate data		
		Cumulative oil production 10^4 t	Cumulative water production 10^4 m^3	Water cut %	Cumulative water production 10^4 m^3	Water cut %	Degree of recovery %
9	1967	581.1	249.2	64.9	269	72	24.12
10	1968	599.9	281.4	63.0	282	62.5	24.90
11	1969	621.2	316.6	62.3	316	65	25.79
12	1970	644.1	360.7	65.8	363	68	26.74
13	1971	666.3	407.7	68.0	417	71	27.66
14	1972	687.6	460.4	71.3	468	73	28.54
15	1973	709.4	527.6	75.5	525	76.6	29.44
16	1974	730.1	600.3	77.8	600	78	30.30
17	1975	751.9	696.9	81.6	693	80	31.21
18	1976	762.6	729.1	83.5	725	81	31.65

Solution: (1) According to the cumulative oil production and cumulative water cut, the water-drive curve is drawn on semi-logarithmic paper as shown in Figure 4 – 13.

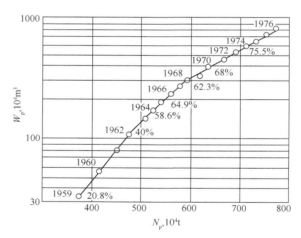

Figure 4 – 13 The curve of cumulative water production and cumulative oil production

(2) Determine the constant sum of the straight line segments and give the equation of the line.

As can be seen from the figure, the straight line has an inflection point in 1968, so it needs to be calculated in stages.

Before 1968,

$$\frac{1}{a_1} = \frac{\lg 1000 - \lg 100}{720 - 480} = \frac{1}{240}$$

Chapter 4 Empirical Methods for Dynamic Analysis

$$a_1 = 240$$

Through the picture $b_1 = 1.023$, $\lg b_1 = 0.01$, therefore the first straight line equation is

$$N_p = 240(\lg W_p - 0.01)$$

Since 1968,

$$\frac{1}{a_2} = \frac{\lg 1000 - \lg 100}{840 - 450} = \frac{1}{390}$$

$$a_2 = 390$$

By the figure, $b_2 = 8.185$, $\lg b_2 = 0.913$, Therefore the second straight line equation is

$$N_p = 390(\lg W_p - 0.913)$$

(3) Calculate cumulative water production W_p and water cut f_w.

The first straight line is taken in 1959,

$$W_p = b \cdot 10^{(N_p/240)} = 1.023 \times 10^{(388.7/240)} = 42.7(10^4 \text{m}^3)$$

$$f_w = \frac{2.3 W_p}{a + 2.3 W_p} = \frac{2.3 \times 42.7}{240 + 2.3 \times 42.7} = 0.29$$

The second straight line is taken in 1975,

$$W_p = 10^{(\lg b_2 + N_p/a_2)} = 10^{(0.913 + 751.9/390)} = 693(10^4 \text{m}^3)$$

$$f_w = \frac{2.3 \times 693}{390 + 2.3 \times 693} = 0.80$$

The all calculation results are shown in Table 4-3.

(4) Calculate the maximum cumulative oil production and ultimate recovery based on the straight line segment.

$$N_{p\max} = 390[\lg(21.3 \times 390) - 0.913] = 1172.5(10^4 \text{t})$$

$$R_{\max} = \frac{N_{p\max}}{N} = \frac{1172.5}{2409} = 49\%$$

The ultimate recovery can also be calculated directly according to equation (4-31).

$$R_{\max} = \frac{390}{2409}\left[\lg\left(21.3 \times \frac{390}{2409}\right) - \lg\frac{8.185}{2409}\right] = 49\%$$

Section 3　Joint Solution of Type B Water Drive Law Curve and Weibull Model

1. Establishment of Weibull Prediction Model

The statistical distribution model proposed by Weibull in 1939 has become the basis of life test and reliability theory research. The distribution density of the model is expressed as

$$f(x) = \frac{\alpha}{\beta} x^{\alpha-1} e^{-(x^\alpha/\beta)} \quad (4-32)$$

where　$f(x)$——Distribution density function of Weibull distribution;

x——Distributed variable, according to the actual problem, the distribution interval is $0 \sim \infty$;

α——Control the shape parameters of the distribution form;

β——Control the scale parameters of distribution peak position and peak value.

If equation (4-32) is integrated, in the interval where x is $0 \sim \infty$, the distribution function value of Weibull can be obtained to be equal to 1. The deduction is as follows:

$$F(x) = \int_0^\infty f(x)\,dx = \int_0^\infty \frac{\alpha}{\beta} x^{\alpha-1} e^{-(x^\alpha/\beta)}\,dx = -\int_0^\infty e^{-(x^\alpha/\beta)}\,d(-x^\alpha/\beta) \\ = -e^{-(x^\alpha/\beta)} \Big|_0^\infty = 1 \quad (4-33)$$

In order to apply Weibull distribution model to the prediction of oil and gas field development indicators, equation (4-32) is rewritten as follows:

$$Q = \frac{C\alpha}{\beta} t^{\alpha-1} e^{-(t^\alpha/\beta)} \quad (4-34)$$

where　Q——annual production of oil and gas fields, $10^4 t/a$ (oil) or $10^8 m^3/a$ (gas);

t——development time of oil and gas field, a;

C——model conversion constant converted from Weibull distribution model to practical model for oil and gas field development.

The expression of cumulative production of oil and gas fields is

$$N_p = \int_0^t Q\,dt \quad (4-35)$$

where　N_p——cumulative production of the oil and gas field, $10^4 t$ or $10^8 t$ (oil), $10^8 m$ (gas).

Substituting equation (4-34) into equation (4-35) and considering the variable transformation method in equation (4-33), t is integrated from 0 to t.

Chapter 4 Empirical Methods for Dynamic Analysis

$$N_p = C[1 - e^{-(t^\alpha/\beta)}] \tag{4-36}$$

When $t \to \infty$, $e^{-(t^\alpha/\beta)} = 0$, then $N_p = C = N_R$, so equation (4-36) can be rewritten as:

$$N_p = N_R[1 - e^{-(t^\alpha/\beta)}] \tag{4-37}$$

After the above results are obtained, the nature and function of the model conversion constant can be explained as follows: due to the Weibull distribution model, the distribution function $f(x) = 1.0$ in the range of x from 0 to ∞, which is equivalent to the cumulative production of the actually developed oil and gas field in the range of t from 0 to ∞, that is, the recoverable reserves of the oil and gas field. Therefore, in order to obtain the result of equation (4-36), the model conversion constant C must be introduced into equation (4-34). The model conversion constant is the recoverable reserves of oil and gas fields. Therefore, equation (4-34) can be rewritten as:

$$Q = \frac{N_R \alpha}{\beta} t^{\alpha-1} e^{-(t^\alpha/\beta)} \tag{4-38}$$

In order to determine the time when the maximum annual output occurs, the derivative of time t is obtained from equation (4-38):

$$\frac{dQ}{dt} = \frac{N_R \alpha}{\beta} t^{\alpha-2} \left[(\alpha - 1) - \frac{\alpha}{\beta} t^\alpha \right] e^{-(t^\alpha/\beta)} \tag{4-39}$$

When $dQ/dt = 0$, there must be $(\alpha - 1) - \frac{\alpha}{\beta} t^\alpha = 0$, so the time when the maximum annual production occurs (t_m) can be obtained as:

$$t_m = \left[\frac{\beta(\alpha - 1)}{\alpha}\right]^{1/\alpha} \tag{4-40}$$

Substituting equation (4-40) into equation (4-38) to obtain the expression of the maximum annual production (Q_{max}) of the oil and gas field:

$$Q_{max} = N_R \left(\frac{\alpha}{\beta}\right)^{1/\alpha} (\alpha - 1)^{1-1/\alpha} e^{-[(\alpha-1)/\alpha]} \tag{4-41}$$

Substituting equation (4-40) into (4-37) again gives the cumulative production (N_{pm}) at the time when the maximum annual production from the oil and gas field occurs as:

$$N_{pm} = N_R \{1 - e^{-[(\alpha-1)/\alpha]}\} \tag{4-42}$$

The remaining recoverable reserves (N_{RR}) of an oil and gas field are expressed as:

$$N_{RR} = N_R - N_p \tag{4-43}$$

Substituting equation (4-37) into equation (4-43) yields

$$N_{RR} = N_R e^{-(t^\alpha/\beta)} \tag{4-44}$$

The reserve-to-exploitation ratio (ω) of remaining recoverable reserves is expressed as:

$$\omega = N_{RR}/Q \tag{4-45}$$

Substituting equations (4-38) and (4-44) into (4-45) yields

$$\omega = \frac{\beta}{\alpha t^{\alpha-1}} \qquad (4-46)$$

The oil recovery rate of the remaining recoverable reserves is the reciprocal of the storage and recovery ratio, so the expression of the oil recovery rate (v_o) of the remaining recoverable reserves is obtained from equation (4-46):

$$v_o = \frac{\alpha t^{\alpha-1}}{\beta} \qquad (4-47)$$

where, v_o is expressed as a decimal, if it is changed to a percentage (%), equation (4-47) is changed to the following equation:

$$v_o = \frac{100\alpha t^{\alpha-1}}{\beta}\% \qquad (4-48)$$

2. Joint Solution of Type B Water Drive Curve and Weibull Model

The Weibull prediction model was studied and derived using the Weibull distribution in mathematical statistics. The model has the function of predicting field production, cumulative production and recoverable reserves, and its basic relations are

$$Q_o = at^b \exp\left[-\frac{t^{b+1}}{c}\right] \qquad (4-49)$$

$$N_p = \frac{ac}{b+1}\left[1 - \exp\left(-\frac{t^{b+1}}{c}\right)\right] \qquad (4-50)$$

$$N_R = \frac{ac}{b+1} \qquad (4-51)$$

Type B water drive law curve was firstly proposed in 1978 by Mr. Tong Xianzhang, a famous expert in China, in the form of an empirical equation. Its theoretical derivation was completed by the literature, and its basic relationship equation is

$$\lg L_p = A + BN_p \qquad (4-52)$$

The derivative of equation (4-52) with respect to time t yields

$$\frac{1}{2.303 L_p} \cdot \frac{dL_p}{dt} = B\frac{dN_p}{dt} \qquad (4-53)$$

It is known that

$$\frac{dL_p}{dt} = Q_o + Q_w \qquad \frac{dN_p}{dt} = Q_o \qquad \frac{Q_w}{Q_o} = R_{wo}$$

Therefore, from equation (4-53), we have

$$L_p = \frac{1}{2.303B}(1 + R_{wo}) \qquad (4-54)$$

Chapter 4 Empirical Methods for Dynamic Analysis

Substitute equation (4 – 54) into equation (4 – 52) to obtain

$$\lg(1 + R_{ow}) = A + BN_p + \lg 2.303B \tag{4-55}$$

Taking the economic limit water – oil ratio $(R_{wo})_L$, the relationship equation for predicting the recoverable reserves of the oil field is obtained from equation (4 – 55):

$$N_R = \frac{\lg[1 + (R_{wo})_L] - (A + \lg 2.303B)}{B} \tag{4-56}$$

The relationship between water-oil ratio and water cut is known to be

$$R_{wo} = \frac{f_w}{1 - f_w} \tag{4-57}$$

Substitute equation (4 – 57) into equation (4 – 55) to obtain

$$f_w = 1 - 10^{-(A + BN_p + \lg 2.303B)} \tag{4-58}$$

Substitute equation (4 – 50) into equation (4 – 58) to obtain

$$f_w = 1 - 10^{-\left(A + B\left\{\frac{ac}{b+1}\left[1 - \exp\left(-\frac{t^{b+1}}{c}\right)\right]\right\} + \lg 2.303B\right)} \tag{4-59}$$

Once the predicted oil production and water cut are obtained from equations (4 – 49) and (4 – 59), the water production and liquid production of the field can be predicted from the following equations, respectively:

$$Q_w = Q_o \left(\frac{f_w}{1 - f_w}\right) \tag{4-60}$$

$$Q_L = Q_o \left(\frac{1}{1 - f_w}\right) \tag{4-61}$$

The equation for the time t_m at which the maximum annual production occurs is as follows.

when

$$\frac{dQ_o}{dt} = at^{b-1} e^{-\frac{t^{b+1}}{c}} \left(b - \frac{b+1}{c} t^{b+1}\right) = 0$$

that is when

$$b - \frac{b+1}{c} t^{b+1} = 0 \qquad t_m = \left(\frac{bc}{b+1}\right)^{\frac{1}{b+1}}$$

The maximum annual production Q_{max} is

$$Q_{max} = a \left(\frac{bc}{b+1}\right)^{\frac{b}{b+1}} e^{-\frac{b}{b+1}}$$

3. Solution Method of the Joint Model

In order to determine the model constants a, b, and c of the prediction model and the value of the recoverable reserves N_R, equation (4 – 49) can be treated as follows:

$$\lg \frac{Q_o}{t^b} = \lg a - \frac{1}{2.303c} t^{b+1} \tag{4-62}$$

If set

$$\alpha = \lg a \tag{4-63}$$

$$\beta = \frac{1}{2.303c} \tag{4-64}$$

Then we get

$$\lg \frac{Q_o}{t^b} = \alpha - \beta t^{b+1} \tag{4-65}$$

Based on the actual development data, the linear trial difference was first solved using equation (4-65) to find b based on the maximum linear correlation coefficient, and then α and β were obtained using the least squares method. The following equations, rewritten from equations (4-63) and (4-64), were then used to determine the values of the model constants a and c, respectively:

$$a = 10^\alpha \tag{4-66}$$

$$c = \frac{1}{2.303\beta} \tag{4-67}$$

After determining the prediction model parameters a, b and c, the recoverable reserves N_R can be solved according to equation (4-51).

In determining the prediction model constants a, b, c and N_R, whether their values are correct and reliable is determined by comparing the theoretical oil production, cumulative oil production and water cut predicted by using equations (4-49), (4-50) and (4-59) with the actual oil production, cumulative oil production and water cut, and the parameters that achieve the best fit are the most accurate and reliable.

4. Practical Application

[**Example 4-3**]   The development data of an oil field is shown in Table 4-4.

Table 4-4   Development data of an oil field

Year	Time, a	Q_o, 10^4t/a	Q_w, 10^4t/a	N_p, 10^4t	W_p, 10^4t	L_p, 10^4t
1965	1	106.03	7.74	106.03	7.74	113.77
1966	2	63.39	8.52	169.42	16.26	185.68
1967	3	48.42	11.78	217.84	28.04	245.88
1968	4	68.62	1.68	286.46	29.72	316.18
1969	5	81.37	3.86	367.83	33.58	401.41
1970	6	92.08	6.58	459.91	40.16	500.07
1971	7	109.68	13.34	569.59	53.49	623.08

continued

Year	Time, a	Q_o, 10^4 t/a	Q_w, 10^4 t/a	N_p, 10^4 t	W_p, 10^4 t	L_p, 10^4 t
1972	8	123.51	21.89	693.10	75.38	768.48
1973	9	121.73	23.92	814.83	99.30	914.13
1974	10	146.50	36.10	961.33	135.40	1096.73
1975	11	173.16	57.28	1134.49	192.68	1327.17
1976	12	180.14	93.50	1314.63	286.18	1600.81
1977	13	171.21	133.54	1486.35	419.71	1906.06
1978	14	163.19	175.50	1649.54	595.21	2244.75
1979	15	161.12	209.52	1810.66	804.73	2615.39
1980	16	151.50	273.69	1962.16	1078.42	3040.58
1981	17	145.47	347.11	2107.63	1425.53	3533.16
1982	18	130.19	391.83	2237.82	1817.36	4055.18
1983	19	116.26	439.37	2354.08	2256.73	4610.81
1984	20	99.20	436.81	2453.28	2693.59	5146.87
1985	21	89.94	471.18	2543.22	3164.77	5707.99
1986	22	79.87	470.17	2623.12	3634.94	6258.06
1987	23	72.00	491.61	2695.12	4126.55	6821.67

Solution: (1) Linear regression based on actual oil and gas field production data to find the intercept and slope of the B-type water drive curve.

The cumulative liquid production (L_p) and corresponding cumulative oil production (N_p) data in Table 4-4 were plotted in Figure 4-14 according to the linear relationship of equation (4-52), and a good straight line was obtained. The intercept of the straight line $A = 2.60$; the slope of the straight line $B = 0.000454$; the correlation coefficient of the straight line $r = 0.9997$ was obtained by linear regression.

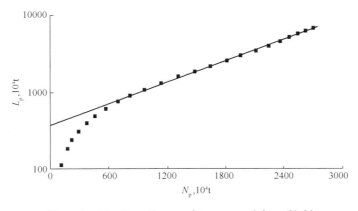

Figure 4-14 Type B water drive curve of this oilfield

(2) Determine the Weibull model constants a, b and c.

The corresponding development data of Q_o and t within Table 4-4 were subjected to a linear trial difference according to equation (4-65), and b was found according to the maximum correlation coefficient (Figure 4-15), and then a and c were solved to obtain: $r = 0.999$; $a = 19.99$; $b = 1.100$; $c = 328.10$. Then, the values of a, b, and c were substituted into equation (4-51) to obtain: $N_R = 3178.27 \times 10^4 \text{t}$.

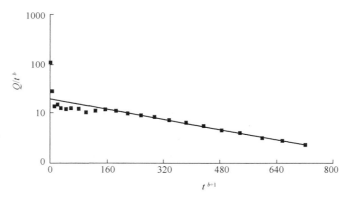

Figure 4-15 Semi-logarithmic plot of Q_o/t^b versus t^{b+1}

(3) Calculate the annual oil production, cumulative oil production, water cut and recovery.

By substituting the values of a, b and c, into equations (4-49) and (4-50) respectively, the relevant equations for predicting the theoretical and cumulative oil production of the field can be obtained as:

$$Q_o = 19.9 \times t^{1.1} \exp\left(-\frac{t^{1.1+1}}{328.10}\right) \qquad (4-68)$$

$$N_p = \frac{19.9 \times 328.10}{1.10+1}\left[1 - \exp\left(-\frac{t^{1.1+1}}{328.10}\right)\right] \qquad (4-69)$$

The values of A, B, a, b and c are then substituted into equation (4-59) to obtain the relevant equation for predicting the water cut of the field as:

$$f_w = 1 - 10^{-\left(2.6+0.000454 \times \left\{\frac{19.99 \times 328.10}{1.10+1}\left[1-\exp\left(-\frac{t^{1.1+1}}{328.10}\right)\right]\right\}+\lg(2.303 \times 0.000454)\right)} \qquad (4-70)$$

If the values of A, B and N_R obtained above are substituted into equation (4-58), the ultimate water cut of the field under abandoned conditions can be obtained as:

$$f_w = 1 - 10^{-[2.6+0.000454 \times 3178.27 + \lg(2.303 \times 0.000454)]}$$
$$= 0.913 (\text{or } 91.3\%)$$

(4) Calculation of the maximum annual production.

The maximum annual production occurs when t_m is

$$t_m = \left(\frac{bc}{b+1}\right)^{\frac{1}{b+1}} = 11.6(a)$$

The maximum annual production Q_{max} is

$$Q_{max} = a\left(\frac{bc}{b+1}\right)^{\frac{b}{b+1}} e^{-\frac{b}{b+1}} = 175.44(\times 10^4 t)$$

(5) Plotting actual annual field production against forecast production and actual cumulative production against forecast cumulative production.

When given different development times t, the predicted values of theoretical oil production Q_o, cumulative oil production N_p and water cut f_w for the field are obtained from equations (4-68), (4-69) and (4-70), listed in Table 4-5 and plotted in Figure 4-16 to Figure 4-18.

Table 4-5 Comparison of actual and predicted data

Time, a	Q_o, 10^4 t/a		N_p, 10^4 t		f_w, %	
	actual value	predicted value	actual value	predicted value	actual value	predicted value
1	106.03	19.93	106.03	9.50	6.8	0.00
2	63.39	42.29	169.42	40.54	11.8	0.00
3	48.42	64.92	217.84	94.17	19.6	0.00
4	68.62	86.85	286.46	170.14	2.4	0.00
5	81.37	107.35	367.83	267.39	4.5	0.00
6	92.08	125.83	459.91	384.17	6.7	0.00
7	109.68	141.78	569.59	518.20	7.3	0.00
8	123.51	154.85	693.10	666.77	15.1	0.00
9	121.73	164.78	814.83	826.86	16.4	0.00
10	146.50	171.46	961.33	995.26	19.8	15.12
11	173.16	174.90	1134.49	1168.70	24.9	29.20
12	180.14	175.20	1314.63	1344.00	34.2	41.05
13	171.21	172.60	1486.35	1518.14	43.7	50.86
14	163.19	167.40	1649.54	1688.34	51.8	58.87
15	161.12	159.98	1810.66	1852.20	56.5	65.35
16	151.50	150.74	1962.16	2007.69	64.4	70.54
17	145.47	140.11	2107.63	2153.21	70.5	74.70
18	130.19	128.53	2237.82	2287.59	75.1	78.02
19	116.26	116.41	2354.08	2410.10	80.0	80.66
20	99.20	104.12	2453.28	2520.36	83.1	82.77
21	89.94	92.00	2543.22	2618.40	84.0	84.44
22	79.87	80.31	2623.12	2704.50	85.5	85.78
23	72.00	69.28	2695.12	2779.24	87.2	86.85
24	—	59.07	—	2843.34	—	87.70
25	—	49.79	—	2897.69	—	88.38

continued

Time, a	Q_o, 10^4 t/a		N_p, 10^4 t		f_w, %	
	actual value	predicted value	actual value	predicted value	actual value	predicted value
26	-	41.48	-	2943.24	-	88.92
27	-	34.17	-	2980.99	-	89.35
28	-	27.83	-	3011.91	-	89.69
29	-	22.42	-	3036.96	-	89.96
30	-	17.85	-	3057.03	-	90.17
31	-	14.06	-	3072.92	-	90.33
32	-	10.95	-	3085.38	-	90.45
33	-	8.44	-	3095.03	-	90.55
34	-	6.43	-	3102.43	-	90.62
35	-	4.85	-	3108.04	-	90.68
36	-	3.61	-	3112.24	-	90.72

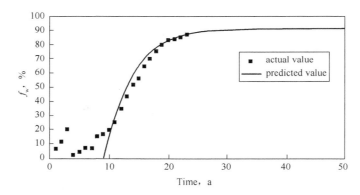

Figure 4-16 Comparison curves between actual and predicted water cut

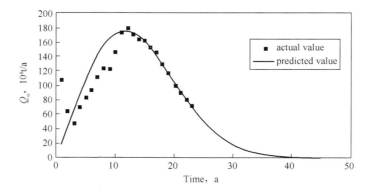

Figure 4-17 Comparison curves of actual and predicted annual oil production

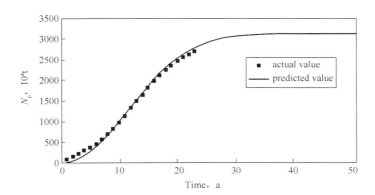

Figure 4 – 18　Comparison curves of actual and predicted cumulative oil production

Chapter 5 Theory and Technology for Oilfield Development Adjustment

After implementation of the oilfield development plan, it is a process with continuous change and corresponding adjustment. During the development of an oilfield, the conditions of exploitation and production gradually change, and the understanding of the reservoir continues being enriched and deepened, these make it necessary to adjust the development deployment accordingly.

Section 1 Monitoring, Analysis and Evaluation of Informational Data in Oilfield Development

1. Monitoring of Information in Oilfield Development

The so-called development monitoring is the real-time recording of various parameters in the oilfield development process. The purpose of development monitoring is to provide data parameters for development analysis and evaluation. The main contents of reservoir engineering monitoring include oil, gas and water production rate, formation pressure, oil (liquid) production profile and water injection profile, produced fluid properties, oil-gas-water contact and its movement, water flooding front advancing law, formation water saturation variation, sand production, and connectivity between wells, etc.. The monitoring results are organized into a series of charts and curves which can be easily used for analysis, comparison and evaluation. Formation pressure is monitored through periodic pressure buildup well tests.

The monitoring of liquid production profile or oil production profile can only be completed through regular production logging. There are many types of production logging, such as flowmeter, resistivity, and radioisotope, etc.. The interpretation results of production logging can be illustrated in the form of Figure 5 – 1. The figure shows that the oil well has two pay zones, and the upper one is the main oil producing zone. The liquid production profile a year ago shows that the upper layer produces oil, and the lower production layer produces liquid with 50% water. The liquid production profile one year later shows that the upper production layer starts to produce water, and the lower production layer has been 100% water flooded. Immediate measures should be taken to seal the lower production layer completely.

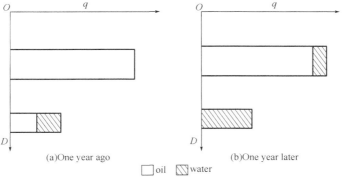

Figure 5-1 Fluid production profile of oil well

Monitoring of the water absorption profile of water injection well is very similar to that of the liquid production profile of oil production well. According to the monitoring results of water absorption profile, corresponding profile control and water blocking measures can be formulated.

Monitoring of formation water saturation can be accomplished by a regular individual-well or inter-well tracer tests, or by drilling specialized inspection wells. Inter-well tracer monitoring can simultaneously detect reservoir connectivity between wells.

Monitoring the oil-gas-water contact or the movement law of the water drive front requires special observation wells. By regularly measuring the position of the oil-water contact and the water flooding front through the observation wells, the movement of the oil-water contact and the water flooding front can be located (Figure 5-2 and Figure 5-3). At the same time, crucial study can also be carried out by numerical simulation method.

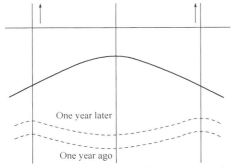

Figure 5-2 Movement of oil-water contact in observation wells

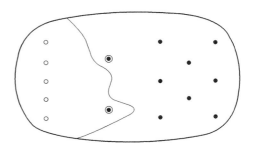

Figure 5-3 Movement of water flooding front in observation wells

The scope and content of development monitoring is fairly extensive, and the monitoring content and monitoring plan should be designed according to the specific needs of oilfield development practice. The monitoring results directly reflect the formation characteristics of the reservoir and the correctness of the development measures.

2. Analysis and Evaluation of Oilfield Development Practice

The so-called oilfield development analysis and evaluation are to use certain methods and means to study, analyze and evaluate various dynamic data obtained in the process of oilfield development, gradually improve the understanding of the geological characteristics of oil reservoir and the regularity of oil and gas movement, revise the geological model continuously, and adjust the oilfield development plan in time, so that the oilfield development work can be sustainably adjusted and run under strict controlling.

2.1 Re-understanding of geological features

In the stage of reservoir evaluation, the establishment of a reservoir geological model is carried out through many indirect data and very limited drilling and test production data, so the understanding about the geological characteristics of oil reservoirs is relatively incomplete. As the development process continues, a lot of static and dynamic data respecting the reservoir are additionally collected. In order to make the geological model of oil reservoirs closer to the real situation, it is necessary to re-study the geological characteristics of oil reservoirs, correct the previous misunderstandings due to limited data, and supplement the gaps in the previous understandings. The methods and contents of the re-recognition of geological features are the same as the work in the oil reservoir evaluation stage. It's just that the re-understanding of geological features is a research work that is regularly or irregularly repeated and unstopped throughout the whole life of oilfield.

2.2 Re-estimation of reserves and recoverable reserves

Due to the enrichment of new data and information, the knowledge of the geological characteristics of oil reservoir changes, in some case, significantly, and the geological model of oil reservoir will be further modified. Since reserves and recoverable reserves are the material basis for deciding the development principle to follow, it is compulsory to recalculate reserves and verify the progress of recoverable reserves after amending the oil reservoir geological model. The geological model of oil reservoir is constantly improved during the development process, therefore, the recalculation of reserves and recoverable reserves must also be repeated year by year. Due to the continuous increasing of dynamic data in the development process, in addition to the conventional volumetric method, some dynamic approaches, such as material balance equation, water flooding curve and numerical simulation, are often used in reserve re-evaluation work.

2.3 Analysis of reserve employment status

It is impossible for a certain development system to employ all the geological reserves. One of the tasks of reservoir engineers is to apply effective means and measures to produce the geological reserves as much as possible. The higher the degree of reserves employed, the more oil produced. Therefore, in the oilfield development practice, it is of great importance to analyze the production status of oil reserves and investigate the measures to enhance the employment degree of reserves.

Water absorption profile and liquid producing profile, sealed coring, separate zone test, and individual layer production data are commonly used in field practice to diagnose and determine the vertical sweep efficiency of the injected agent, the watered out and water washed situation, and the oil and liquid production from every pay zone. Field test data and reservoir engineering analysis are frequently adopted to obtain the areal sweep efficiency and oil displacement efficiency. The trend of employment of reserves can be predicted by using water flooding curve. The distribution of remaining oil in the reservoir can often be described by using material balance equation and numerical simulation, etc..

2.4 Analysis of production rate

The core of oilfield development is oil production. The magnitude of oil production rate and its composition reflect level and stage of oilfield development. Oil production has become one of the main objects of oilfield dynamic analysis. Usually, oil production can be observed from three sets of curves: (1) oil production composition curves, including daily production rate of old wells, daily production rate of stimulated wells and daily production rate of newly drilled wells, (2) overall production curves, including daily oil production rate, daily liquid production rate, daily injection rate, comprehensive water cut and comprehensive gas-oil ratio, and (3) decline curves, including natural and comprehensive decline rate.

Through the analysis and investigation of the above three sets of curves, the current status of oilfield can be estimated, the future trend of oilfield can also be predicted, and suggestion with regard to stimulating measures effective to slow down and stabilize oil production rate can be put forward.

2.5 Analysis of injection production balance

The balance of injection and production is a significant guarantee for maintaining the energy of oil reservoir and achieving long-term high and stable production. During the development process, it is required to timely analyze the balance of injection and production of oil reservoir. The main tasks of injection-production balance analysis are:

(1) The relationship between the injection-production ratio and the pressure maintenance level, and the rationality of the pressure system and the ratio of injection-production wells.

(2) Reasonable pressure maintenance level, suitable energy efficiency, analysis of reservoir depletion, and evaluation of the effect of production and injection plan.

(3) Appropriate pressure profile, water injection pressure difference and oil production pressure difference, during different development stages.

2.6 Stimulation Measures for wells and pay zones

In order to wisely exploit oil resources and make full use of driving energy, various measures are frequently conducted to stimulate oil production wells, water injection wells and reservoir zones, during the development process. Before implementing stimulation measures, feasibility analysis should be carried out. After the measures are executed, the effect must be concluded, such as the quantity of incremental of oil production and water injection, the validity period of the measures, and the influence on constant oil production and decline rate.

2.7 Analysis of remaining oil

Remaining oil refers to the crude oil remaining underground that has not been produced until a certain time point. The mechanism of remaining oil mainly includes three aspects: (1) not swept, (2) not fully swept, and (3) insufficient displacement efficiency. "Remaining oil" is the heart of oilfield development, involving two key jobs, namely the description of the remaining oil and the tapping of the remaining oil. The description of remaining oil is the fundamental for the potential tapping.

The content of the remaining oil description involves: (1) the location of the remaining oil, (2) the amount of the remaining oil, such as reserves, reserves abundance, unit reserve factor, and remaining oil saturation, etc., and (3) the mechanism binding the remaining oil, that is, the causes resulting in the remaining oil.

2.8 Overall evaluation of oilfield development effect

Analysis of oilfield development process is an extremely complex task. It is by no means analyzing only a single index alone or all indexes but isolated from each other. It needs to take into account all aspects regarding oilfield development comprehensively, and thus, a systematic idea effective to guide the whole development process. Generally, oilfield development analysis requires performing the following tasks.

(1) Establish oilfield development database: record the static and dynamic data of oilfield and provide convenience for fully sharing the data resources.

(2) Analyze the adaptability of the injection production system: whether the injection production system designed in the oilfield development plan is suitable for the actual needs, whether it is suitable for all of the high, medium and low water cut stages throughout the development process, whether it is suitable for various oil production technologies such as natural flow, pumping and gas lift, etc., whether the injection production system needs to be adjusted, and, whether the driving energy is sufficient, etc..

(3) Analyze the potential of pay zones: the degree of reserves employment, the distribution of remaining oil, whether the displacement mechanisms are appropriate, and new measures to improve recoverable reserves and recovery factor.

(4) Three laws are analyzed: oil production declining, water cut rising and pressure changing. Through the analysis of these three laws, we have a macroscopic image on oilfield performance, and then, according to the problems reflected by the three laws, we propose corresponding countermeasures.

(5) Analysis of the effect of stimulation measures and development adjustment: whether the adjustments are valid and the degree of the effects, including, groups of development zones, well pattern, injection production system, driving mechanisms, injection and production rate, and stimulation measures such as fracturing and acidizing.

(6) Economics analysis: how about the economic benefit of the development system adopted, whether the economic indexes such as oil production cost and investment return rate are optimal and whether the development system needs to be greatly adjusted, etc..

Section 2   Content of Oilfield Development Adjustment

Based on the monitoring and analysis during oilfield development practice, the original design of oilfield development plan is comprehensively evaluated. In general, early development plan designed with less information is to more or less extent ill-suited to reservoir realities. In order to improve the development effect of oilfield, it is necessary to modify the original design appropriately, that is, according to the new geological understanding and the current economic and technical conditions, a new set of development plan, called the adjustment plan, is worked out. The preparation method of the adjustment plan is basically the same as that of the initial plan.

The content of adjustment in oilfield development process includes: (1) adjustment of groups of development zones, (2) well pattern adjustment, covering well density, injection production system and well pattern shape, etc., (3) adjustment of driving mechanism, and (4) adjustment of production technology.

Oilfield development adjustment can be global or local, and in most cases, only local adjustment is required.

The adjustment of group of development zones is generally a finer dividing of groups of zones, based on the new conception of geological data. Original dividing of groups of development layers, after a certain of period of running, exhibits many inconsistencies between different zones. The high permeability zones have been severely water flooded, but, the crude oil in the low permeability zone is still not swept or just employed to a very low degree and becomes remaining oil. Aiming at this problem, the development groups should be further refined to increase the degree of oil recovery (Figure 5 – 4). Adjustment of groups of development zones generally adheres to the principle of "from coarsely dividing turning towards finely dividing".

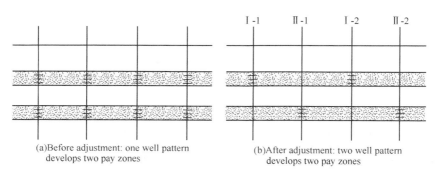

(a) Before adjustment: one well pattern develops two pay zones

(b) After adjustment: two well pattern develops two pay zones

Figure 5 – 4 Finely dividing of the groups of development layers

The adjustment of the development well pattern is to infill the well pattern and modify the injection production well system, based on the new geological understanding, The original well pattern, after a certain period of production, shows some inadaptability, mainly manifesting as too large well spacing resulting in lower degree of covering of sand bodies or reserves. In this case, the development well pattern should be further infilled to improve the degree of crude oil recovery (Figure 5 – 5). In order to enhance oil production rate, the form of well pattern is generally revised from high to low ratio of injection to production wells (Figure 5 – 6). Well pattern adjustment generally adheres to the principle of "from a thinner one turning towards a denser one".

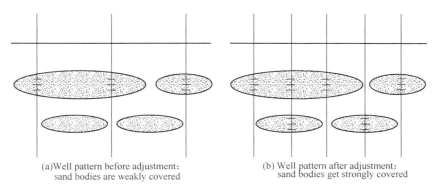

(a) Well pattern before adjustment: sand bodies are weakly covered

(b) Well pattern after adjustment: sand bodies get strongly covered

Figure 5-5　infilling of the development well pattern

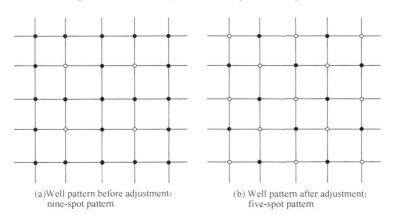

(a) Well pattern before adjustment: nine-spot pattern

(b) Well pattern after adjustment: five-spot pattern

Figure 5-6　Adjustment of the development well pattern

● water injection well　○ production well

The adjustment of driving mechanism generally refers to the changing from depleted drive to water (or gas) injection, or from water flooding to EOR procedures, based on new geological realization. The essential principle is to modify the driving mechanism from one of lower displacement efficiency to one of higher efficiency.

Small scale adjustments are carried out at any time during the development, such as the optimization of the working parameters of wells, the variation of fluid flow direction, the alteration of perforations, and the EOR technologies for tapping remaining oil such as local well infilling, turning some oil wells into water injection, profile modification, water plugging and fracturing. Major adjustments are usually taken at a time interval of more than 3~5 years, such as the adjustments of group of development layers, driving mechanism, and development well pattern. Oilfield development adjustment is a never-ending job.

Development adjustment is generally designed according to the needs of oil and gas production, with the principle of increasing oil and gas production and improving economical benefit.

The fundamental purpose of development adjustment is to change the field of porous flow through various technological means so that the static remaining oil flows, the slowly moving oil flows faster, the affected area of injected water gradually expands, the dominant low-efficiency flowing passages are blocked, and thus the displacement efficiency by injected water can be significantly enhanced.

Section 3　Experience and Prospect of Development Adjustment in Lamadian, Saertu, and Xingshugang Oilfields

Diversity, complexity, concealment, and time variability determine that oilfield development must go through the process of many times of adjustment, and each time of adjustment should establish a development mode conforming to the specific situation of this oilfield and this stage. Because of the coexistence of multiple well patterns, various driving mechanisms and plugging of some layers, the establishment of development modes for Lamadian, Saertu and Xingshugang oilfields are more complicated.

1. Five Types of Development Modes

Since put into development in 1960, Lamadian, Saertu and Xingshugang oilfields have experienced the adjustments including primary well pattern, the first round infilling well pattern, the second round infilling well pattern, the third round infilling well pattern, alterations of driving mechanism, changes of injection production system, modifications of injection production composition, more finely dividing water injection intervals, polymer flooding, and the industrial applications of improving oil recovery technologies, and have established five representative development modes. They are (1) executing early stage internal transverse division line well pattern water flooding to maintain formation pressure, (2) employing reserves step by step to keep stable production consecutively, (3) modifying the production composition to sustain stable oil output rate and suppress water production, (4) "54321" fine development adjustment, namely, five "not equal to", four "fine", three "level", two "control" and one "new way to finely and efficiently develop oilfield at its ultra-high water cut period", and (5) parallel applications of polymer flooding and water flooding.

Chapter 5 Theory and Technology for Oilfield Development Adjustment

1.1 Early stage internal transverse division line well pattern water flooding to maintain formation pressure

In the early stage of oilfield development, the working conditions were difficult, but the famous ten pilot tests of oilfield development technologies were still carried out. Based on the pilot tests, strategic investigation and referencing to domestic oilfield development experiences, taking into account the specific characteristics of Lamadian, Saertu and Xingshugang oilfields, the development mode of "early stage internal transverse division line well pattern water flooding to maintain formation pressure" was innovatively put forward.

(1) Early stage water injection. Take water injection served as the key to maintain formation pressure, so that oil wells maintained rigorous production capacity for a long time. The injection production ratio was kept at around 1 and the formation pressure was remained near the original formation pressure.

(2) Line well pattern. The oilfield was developed with line well pattern of lager transverse divisional distance. For anyone of the transverse divisions, three rows of oil wells are arranged between two rows of water injection wells, the distance from the first row of oil wells to the row of injection wells is 600m, the distances between the rows of oil wells are all 500m, and the well spacing in all rows is 500m. Oilfield areas put into development after that time modified the well spacing and row spacing to some extent.

(3) The PI reservoirs are of the best properties and thus were taken into the same one independent group of development zones. The other reservoirs were taken as another one group. In the process of implementation, some oilfield areas conducted certain adjustments.

(4) Lamadian oilfield and Gaotaizi reservoirs were taken as backup reserves, did not be produced and were waiting for development at appropriate time.

(5) Keep the formation pressure constant to support flowing production. The bottomhole pressure of production wells was allowed to be slightly, 0.5 ~ 1MPa, lower than saturation pressure.

(6) Based on the reasonable division of groups of development layers, total water injection in one well should be intently allocated to every layer according to its specific need and capability, so that advance of water front was controlled uniform.

(7) Oil wells near the rows of water injection wells were firstly put into production, leaving the middle row of oil wells as a warehouse, did not be produced. After the two side rows of oil wells produced for 5 years and saw water breakthrough, the middle row of oil wells were put into production. Then, a stable oil production velocity of about 3% was held for 10 years.

(8) Do the best to extract more crude oil during water free stage. There was basically no water be produced within 5 years. However, we should make full preparations for the long term production of water bearing crude oil.

(9) The research and innovation about technologies such as selective water injection, selective water plugging and selective fracturing, were concerned and reinforced.

This development mode was creative and effectively guided the overall development of Lamadian, Saertu and Xingshugang oilfields. Since 1960, the annual oil production rate went up sharply year by year. By 1976, when the whole oilfield was fully put into development, the annual oil production reached a high peak of more than 5000×10^4 t.

1.2 Employ reserves step by step to keep stable oil production consecutively

In 1976, the Lamadian, Saertu and Xingshugang oilfields were totally put into development, and the decision makers put forward the goal of "achieving a high production rate up to 5000×10^4 t/a and keeping it stable for at least ten years". There was a great debate about this goal, because no any oilfield throughout the world has ever maintained that a high stable production rate for 10 years. Therefore, extensive and thorough researching investigations were organized and brought into force, and as the result, the development mode of "producing different types of oil layer in different batches respectively and employing reserves consecutively" came into truth.

(1) Relying on four footholds, stable production rate was sustained during period of the Fifth Five-Year Plan. The comprehensive adjustment focusing on technology of six "separate" and four "clear" were reinforced. Four footholds were proposed, including primary well pattern, major oil zones, six "separate" and four "clear", and natural flowing production.

(2) "Three changes" was implemented. The dominant adjustment measure was changed from six "separate" and four "clear" to infilling well pattern. The main adjustment objects were changed from the major oil layers to the non-major oil layers and the thin and poor oil layers. The producing technology was changed from natural flow to artificial lift. The first round of infilling well pattern was completed, the task of which was to finely divide groups of development layers.

(3) The injection production system was modified, as a result, the ratio of controlled reserves was increased, and formation pressure was restored and held constant. Because ratio of number of oil wells to number of water wells was as large as up to $3.5:1$, the effective directions of liquid supply to oil well was less, and the adjustment of areal water injection situation is difficult. Gradually, ratio of number of oil wells to number of water wells was altered to less than $2:1$, and thus the oilfield development status was significantly improved.

(4) The second round of infilling well pattern was actuated, hence, the production degree of thin and poor oil zones and untabulated layers was raised. After the first round infilling well pattern, there is still some small scale poor pay zones difficult to be employed. And moreover, some untabulated zones with relatively better properties, also had oil producing ability. Therefore, aiming to exploit these oil layers, the second round infilling well pattern was put into action.

(5) After the second round infilling well pattern, some regions and oil zones still had remaining potential due to unemployment or poor employment. In addition, untabulated layers had certain potential. Therefore, some oilfield regions had the material basis for further adjustment and the third round infilling well pattern was constructed.

This development mode established the theory and technology of separate zone water injection, affirmed the leading role of Daqing oilfield development in the world, supported the long term high and stable production of Daqing oilfield, and acquired huge economical and social benefits.

1.3　Modify the production composition to sustain stable oil output rate and suppress water production

In 1990s, the comprehensive water cut of Lamadian, Saertu and Xingshugang oilfields had exceeded 80% and the oilfields entered into the later stage of high water cut. The first round infilling well pattern had been put into running, the second round infilling adjustment was under executing, and the annual oil production rate went up to the peak of 5600×10^4 t. From the perspective of overall development potential, the second round infilling well pattern and peripheral oilfields had larger production capacity. To deal with the problem of the rapid increase of fluid production and the obvious difference in development status among various development regions, various wells and various layers, the development mode, modifying the production composition to sustain stable oil output rate and suppress water production, was proposed.

(1) According to the status of various oil wells, various layers and various regions, customized adjustments were made for different types.

(2) The liquid output from layers, oil wells and blocks with high water cut were depressed, the liquid output from layers, oil wells and blocks with lower water cut were enhanced, so the development of poorly employed or unemployed layers was strengthened.

(3) Conventional technologies such as separate zone water injection, hydraulic fracturing, infilling well pattern, injection and production system optimization, and regulating pressure system, were used to modify water injection composition, liquid output composition and reserves production composition.

(4) For the wells, layers and blocks with obvious problems and large differences in development status, mass treatments were carried out, and the experiences were comprehensively concluded, popularized and routinized.

(5) The goal was to control the rise of water cut, control the rapid growth of fluid production rate, and maintain the stable oil production of the oilfield.

The comprehensive adjustment technology of "modifying the production composition to sustain stable oil output rate and suppress water production" was another technological leap after "separate zone water injection" and "finely dividing groups of development layers and correspondingly arranging the infilling well pattern", and it was the enrichment and advancement of the water flooding technology in Lamadian, Saertu and Xingshugang oilfields of Daqing. This development mode pointed out the direction for daily management and comprehensive adjustment for oilfield operation and made the theory, thought and technology system of water flooding for multi-layer sandstone oilfield more perfect.

1.4 The "54321" fine development adjustment mode

After multiple times of adjustments, the remaining oil and development adjustment potential were highly dispersed. Therefore, "54321" fine development mode was initiated and applied, i. e., five "not equal to", four "fine", three "level", two "control" and one "new way to finely and efficiently develop oilfield at its ultra-high water cut period".

(1) The five "not equal to" includes "the whole oilfield being of high water cut does not equal to that every well being of high water cut", "the whole well being of high water cut does not equal to that every layer in this well being of high water cut", "the whole layer being of high water cut does not equal to that every direction in the layer being of high water cut", "the geological work being fine does not equal to that all subsurface potentials being clearly investigated" and "the development adjustment being fine does not equal to that every block, every well and every layer being adjusted to the maximum extent". This viewpoint of potential of the five "not equal to" was applied to recognize the potential for development. The standard for water injection was refined and quantified. Conventional technologies were run coordinately upon wells to tap potentials.

(2) Aiming at the fact that various contradictions appeared and aggravated during development process, the idea of four "fine" for tapping remaining oil potential was proposed and confirmed, they are fine reservoir description, fine adjustment of injection production system, fine adjustment of injection production composition and fine production management. The three dimensional adjustment was carried out, which contains three levels, i. e., adjustment of areal composition, adjustment of

inter layer composition and adjustment of intra layer composition. Three contradictions, areal, inter layer and intra layer contradictions, were conquered. Target of two restrains were attained, i. e., restraining the decline rate of oil production and restraining the rising of water cut.

(3) Conventional technologies were frequently improved and used. The conventional techniques, such as separate zone water injection, fracturing, water plugging, changing pump, completing the injection production correspondence for single sand body, infilling well pattern, adjustment of injection production system and regulating of pressure system, have been finely improved.

(4) Daqing took the example pilot area as the guide, concluded the experience, and promoted it throughout the whole field. Since the Daqing oilfield put forward the four "fine" in 2008, the water cut rising rate was controlled within 0.5%, the natural decline rate and comprehensive decline rate of oil production were controlled within 9% and 6% respectively. Good effect was achieved from adjustments, Daqing oilfield kept stable annual oil production above 5000×10^4t for 27 years and then 4000×10^4t for 14 years.

1.5 Parallel applications of polymer flooding and water flooding

In 1996, the overall application of polymer flooding made the development of Lamadian, Saertu and Xingshugang oilfields become a situation of "coexistence of polymer flooding and water flooding". On the one hand, as polymer flooding technology was constantly promoted, the water flooding reservoirs were consecutively transformed to polymer flooding. On the other hand, water flooding adjustment was refined continuously according to the development mode of "54321". Therefore, during this period, the two development modes of water flooding and polymer flooding were basically in the parallel state. This stage was a transition stage at which both the modes of water flooding and chemical EOR perfectly contributed to the oil production.

(1) The first grade reservoirs (the major oil layers) were mature for polymer flooding technology and could achieve good technical and economical results, so they were preferentially selected to apply polymer flooding tertiary oil recovery. During the application, the technical parameters and supporting measures were further optimized. Then, polymer flooding was extended to the second grade reservoirs.

(2) The thickness of one set of development layers was controlled in the range of 8~10m, the well pattern was specially dominated by polymer flooding, and the perforations for water flooding were plugged off in polymer flooding well pattern. To improve the utilization of equipments such as polymer solution preparation, injection and well pattern, only one group of development layers was

arranged performing polymer flooding, and the other groups were polymer flooded one by one according to the longitudinally upward sequence.

(3) The polymer solution is of larger viscosity and lower injection ability, hence, the five-spot areal well pattern was adopted and the controlling degree of polymer flooding reached more than 70%.

(4) According to reservoir characteristics, the relative molecular weight of polymer, the concentration of solution and the slug size were optimized. The compatibility of polymer solution with reservoir was enhanced. The viscosity of polymer solution was maintained not less than 40mPa · s. Separate zone injection of polymer solution was carried out, and the studying and exploring the possibility of separate quality separate zone injection were conducted.

In 1996, polymer flooding was popularized and applied in comprehensive industrialization. The ultimate recovery factor was increased by more than 10 percentage points compared to water flooding. The oil production more than 1000×10^4 t was maintained for a long time. Polymer flooding has made and is still making a great contribution to the stable production of 5000×10^4 t and 4000×10^4 t for Daqing oilfield.

Lamadian, Saertu and Xingshugang oilfields are the largest base for research and application of chemical flooding. The prevailed application of polymer flooding technology is one of the milestones of oilfield development engineering throughout the whole world, which leads the trend of enhanced oil recovery and makes Lamadian, Saertu and Xingshugang oilfields the most successful model for research and practice of the technologies of separate zone injection and tertiary oil recovery.

2. The Inherent Rule of the Evolution of Development Mode

Applying the dialectical law of the development and change of things to analyze the evolution of the development modes of Lamadian, Saertu and Xingshugang oilfields, we can see their internal relations and provide ideological guide for future research work.

(1) Reservoir has diversity, complexity, concealment and time variability, which determine that the process of recognition and development adjustment of the reservoir can not be completed at one time. Each time of adjustment brings in a large amount of new information and at the same time new dynamic responses appear, these create favorable conditions for deeper insight into the reservoir. Therefore, the adjustment of oilfield development must go through the process of many repeating cycles of understanding and adjustment.

(2) The evolution process of the first four development modes in Lamadian, Saertu and Xingshugang oilfields is the process of deepening the understanding the oilfields and remoulding the conditions of field production, and also the process from coarse to fine of the development objects. The establishment of the first development mode is of pioneering significance, the core of which is "developing a large oilfield". The basic content of this development mode is to consider the oilfield development from the standpoint of whole oilfield and global system. The second development mode was established, as oilfield development being carried forward, when the production conditions of good layers and poor layers exhibited great difference under the same development conditions. The good oil layers continued to be developed with the basic well pattern, while the poor oil zones were divided as another groups of layers and were developed using infilling well patterns. Therefore, the core of this development mode is "treating the two grades of layers differently, producing them separately, and consecutively keeping the stable oil output rate". The third development mode was established under the situation of the coexistence of various well patterns and large compositional differences in the status of various reservoirs. The core is modifying and optimizing the development situation through composition adjustment, three "separate" and one "optimization" and the backward catching up the advanced. The fourth development mode was established after the adjustments of the aforementioned three development modes, when the remaining oil potential of the water flooding oilfields distributed in a very scattered way. The core is finely digging the potential and squeezing out every drop of oil.

(3) The popularization and application of various EOR technologies has changed the main adjustment direction of oilfield development from expanding sweep volume to improving oil displacement efficiency and controlling the ineffective circulation. The adjustment of traditional water flooding development seems to have reached its "end". Although these four development modes are very different, the common point is that the adjustment direction is the same, which is to "expand the sweep volume and tap the remaining oil". In addition to this direction, there are additional two directions to EOR, one is to improve oil displacement efficiency, and the other is to adjust the difference in displacement degree. There is still debate about whether polymer flooding is mainly about expanding the sweep efficiency or improving the oil displacement efficiency. But there is no doubt that binary, ternary and foam composite flooding can improve the oil displacement efficiency and expand the sweep volume as well. Before the ineffective circulation formed, the difference of displacement degree appears as the contradiction between the advanced and the backward. After the oilfield stepped into the ultra-high water cut stage, the ineffective circulation came into shape, the contradiction of displacement degree changed, which became the

dominant influencing factor restricting the development effect. The ineffective circulation is a "cancer", only destructive, no any benefit, must be dealt with carefully. Therefore, the direction of establishing the development mode in the future should be shifted to improving the oil displacement efficiency, controlling the ineffective circulation and narrowing the difference of the displacement degrees, to initiate the evolution process of the new development mode.

(4) The evolution of the development mode strictly puts into operation the philosophical law of "practicing, understanding, repracticing, and reunderstanding" and "summarizing" cycle by cycle, repeatedly. The adjustment of water flooding and the established development modes were from coarse to fine, went through several times of the process of "practice, understanding, repractice and reunderstanding", and then were upgraded to the technology of tertiary oil recovery, starting the new cycle of repeating "practice and understanding" for new process of development adjustment. From the professional characteristics of oilfield development engineering and the past practical experience, after each time of the cycle of "practice and understanding", the technical points only increased but not decreased, the technical difficulty only increased but not decreased and the management points only increased but not decreased. The practices that have well worked in the past should not be abandoned and new technologies and practices are constantly on the rise.

(5) In the new development stage, there will be more problems and the situation will be more complex. During the new stage, multiple oilfields, such as water flooding, polymer flooding, ASP flooding, gas flooding, after polymer flooding and microbial flooding, coexist, and multiple technologies simultaneously work, including management of ineffective circulation, production of untabulated reservoirs, treatment of inefficient wells, and several ways for EOR. In the new development stage, the second, the third, the fourth and even the fifth times oil recovery interlace tightly, each set of development zones interlaces with the others, and each well pattern intertwines with the others. Research and development adjustment will be more difficult, more research powers and development investments are required.

To sum up, the process of oilfield development adjustment is always "first the big then the small, first the fatty then the thin, first coarse then fine, first easy then difficult". The "first the big then the small" and "first the fat then thin" conform to the idea of addressing the principal contradiction. The "first coarse then fine" and "first easy then difficult" conform to the law of technological evolution. When things evolves to its extreme in a certain direction, a new turning point will appear and a new world will be opened up, which completely conforms to the philosophical law of "practice, understanding, repractice and reunderstanding".

3. Future Development Mode for Lamadian, Saertu and Xingshugang Oilfields

The situation and problems in the new development stage are very complicated, so it is necessary to start the research from problems with urgent need to be solved.

3.1 Problems and understanding

(1) Ineffective circulation has become the main contradiction of water flooding development in the ultra-high water cut stage. It costs more, inhibits low permeability pay zones and dilutes effectiveness. Ineffective circulation is the fundamental causes for the aggravation of the three major contradictions, i. e., areal, interlayer and intralayer contradictions. Controlling inefficient circulation will become the main direction of water flooding development adjustment in the future. The contradiction in developing the two grades of reservoirs is increasing, which embodies in four aspects. For the first, the contradiction between layers the remaining oil has been tapped and layers the remaining oil has not been tapped yet, after the implementation of various adjustment measures, becomes more serious. For the second, the contradiction between the good layers (mainly the ineffective circulating channels) and the poor layer, among the employed layers, increases. For the third, there are great differences and contradictions between the untabulated and tabulated layers. For the fourth, there are more polymers retained in low permeability layers, after polymer flooding.

(2) The whole oilfield has entered into the stage of ultra-high water cut. The water cut of all kinds of wells is gradually approaching each other, the average single well production drops down to lower than 2t/d, and the number of inefficient wells increases day by day. The treatment and comprehensive utilization of low efficiency wells become the problem that must be faced at in the future.

(3) The effect of infilling well pattern for producing untabulated layers is getting worse and worse, which shows great contradiction between untabulated and tabulated layers when they are commingledly produced, is of high water cut and low oil production rate, and the number of inefficient wells in which takes account a large ratio. It is necessary to take more intensive technical measures to alleviate the development contradiction. Therefore, it is important to select favorable blocks to construct independent group of development layers or strengthen water injection in a targeted way.

(4) At present, tertiary oil recovery is the main technology, so it is necessary to establish groups of development zones and corresponding well patterns independently, and put into action

step by step, one by one in the vertically upward sequence. The water flooding is in a cooperative position. As the oil layers are gradually transformed to the tertiary oil recovery process, the injection production system of the water flooding well pattern is dismembered and encroached upon by EOR well patterns. It is demanded to study the utilization of water flooding well patterns.

(5) The technology for further enhancing oil recovery after polymer flooding is mature, and the time of popularization and application is an uncertain factor affecting the overall situation. This uncertainty adds the process of technical development and promotion into the adjustment of the groups of development layers and well patterns, and the adjustment of the groups of development layers and well patterns is no longer only a technical problem.

(6) The comprehensive analysis of oilfield dynamic performance and core sample data shows that the remaining movable oil in the water flooding reservoirs can be divided into five categories. The first is the remaining oil in oil layers unswept by water, which accounts for a relatively small proportion. It shows that the areal control degree of the existing well patterns is very high and the development effect is very good. The second category is the remaining movable oil in the untabulated reservoirs, accounting for a medium proportion, which is the main object of development adjustment of water flooding. The third is the movable oil in the unwashed portion of the water swept layers, which is the remaining oil of the most quantity and is the main potential object of tertiary oil recovery. The remaining movable oil in the washed portion of water swept layers can be divided into two parts, one of which is the fourth type of residual oil, the remaining movable oil that cannot be effectively recovered in the water swept layers, which has been neglected because it has been employed, and it is the main potential object to control the invalid circulation. The other part is the fifth category of residual oil, residual recoverable reserves, which can be effectively recovered by the existing well pattern.

3.2 Basic principles for establishing the development mode

In the future development and adjustment practice, due to the coexistence of the secondary, the tertiary, and the fourth times oil recovery, the coexistence of untabulated reserves potential tapping, ineffective circulation treatment and inefficient well enhancement, the uncertainty of the time to promote the technologies suitable after polymer flooding, and also because of the obvious differences in reservoir conditions, development status and potential situation in different regions, it is difficult to establish a unified development adjustment mode. However, there are still many common points can be used as principle guidance.

(1) The arrangement of groups of development layers and corresponding well patterns of

tertiary oil production, running one by one from the bottom in the vertically upward direction, should be planned and positioned in advance.

(2) The untabulated reserves should be taken as the predominant object for water flooding, favorable areas should be chosen and developed independently, and the water injection adjustment should be comprehensively strengthened.

(3) Different feasible plans should be investigated and decided with respect to the time of field scale application of the effective technologies for reservoirs already completed polymer flooding. The groups of development layers and corresponding well patterns should be studied and optimized.

(4) The groups of development layers and corresponding well patterns for water flooding tabulated reserves should be recombined by sections, taking into consideration of the treatment of ineffective circulation, the treatment of inefficient wells, and the tertiary oil recovery.

(5) To alleviate the growing contradictions during development process, we are required implement more powerful technical measures.

3.3 Subjects required to be vitally studied

To realize the above assumption and establish new development modes, we must strengthen the research in the following five aspects. The first is the research on the fine classification of potential for adjustment, the second is to speed up the research on the main technology suitable for reservoirs after polymer flooding, the third is the research on the effective driving mechanisms for untabulated reservoirs, the fourth is the research on the strategies and technologies for the treatment of ineffective circulation, and the fifth is the research on the combination mode of groups of development layers and corresponding well patterns.

4. Recognitions and Suggestions

(1) Reservoir has diversity, complexity, concealment and time variability, which determine that the process of oilfield development must experience multiple cycles of recognition and adjustment, and must be a course of going forward step by step. The localization of the development mode is the basic technical route for the efficient development of oilfields and an important part of reservoir engineering research. Therefore, we must pay attention to strategy research about the whole development process, predict the oilfield development process and reserve capable technologies in advance.

(2) The evolution of development mode of water flooding in Lamadian, Saertu and

Xingshugang oilfields has finished the course of development adjustment process from coarse to fine, the aim of which is "expanding the swept volume and excavating the remaining oil". It is the perfect embodiment of the philosophical law of "practice, understanding, repractice and reunderstanding". The scale of polymer flooding continues to rise, opening a new stage of development and adjustment after the upgrading of the development mode.

(3) According to the oilfield status, three understandings have been obtained on how to establish the future development mode. The first, the adjustment of water flooding will change to the direction of "dealing with the ineffective circulation, regulating the difference in displacement degree and mining the extremely dispersed remaining oil". The second, the new development mode must adopt the technical consideration of "coexistence of multiple driving mechanisms and simultaneously conducting multiple technologies". The third is, the uncertainty of the time waiting for efficient techniques, after polymer flooding, has a great impact on the establishment of a new development mode.

(4) The ineffective circulation of injected water is the predominant contradiction of oilfield development in the ultra-high water cut period, which has exerted three major impacts on oilfield development, i.e., monetary consumption, inhibition of low permeability layers and dilution of the effectiveness. It has the potential to improve oil recovery by nearly 5 percentage points through controlling ineffective circulation and regulating the difference in displacement degree.

(5) The potential of untabulated layers is great. There is a great contradiction between untabulated zones and effective zones, when they are produced together. Therefore, the research on development strategy should be reinforced.

(6) We should fully realize the complexity of oilfield development in the future, strengthen theoretical, strategic and technical research, investigate and establish an optimized development mode, according to the basic law of practice and understanding, and constantly promote the progress of oilfield development theory and technology.

Section 4 Discussion on Revolutionary Technology for Oilfield Development

Oilfield development is a systematic engineering. Only analyzing, designing and adjusting from the perspective of the full history, the global situation, and the whole system, and using the philosophical thinking of dialectics and evolution, can we achieve the best technical, economic, and social results.

1. Downsizing is the Technological Key to Substantially Improve Oil Production Rate and Ultimate Recovery Efficiency

A technology must have two characteristics can it be defined as revolutionary or transformative technology. As we all know, the first industrial revolution is the steam engine, the second industrial revolution is electricity, and the third industrial revolution is the computer. The common points of these three technologies are, for the first, they bring about significant changes. Whether there are electric light and telephone or not, the changes are significant and revolutionary. For the second, they are put into popularized application in a certain large scale. Every household uses electricity and computers, which can be described as revolutionary large-scale applications.

In the history of the petroleum industry, polymer flooding technology can be called an industrial revolutionary technology. Firstly, it has achieved remarkable results, with more than 10 percentage points of enhanced oil recovery in the field applications. Secondly, it has achieved large-scale application in the petroleum exploitation industry.

The technological key of the next revolution in the oil exploitation industry lies in the downsizing. From exploration means, geological research and reservoir description, to information monitoring, static and dynamic analysis, geological model and reservoir numerical simulation, downsizing all the related tools, technologies and engineering, is the key leading to a revolution, and will consequentially result in a breakthrough to oil production rate and recovery efficiency.

2. Preventive Treatment of Disease Can Get Twice the Result With Half the Effort

For a target oilfield, before its construction and development, we clearly know that sooner or later, the oilfield will form inefficient circulation channels of injected water, and the inefficient circulation of energy becomes the crucial problem restricting the effect of oilfield development. Therefore, from the beginning, we should put the problem of inefficient circulation as one of the key points of attention. Every year, we should track, analyze and evaluate the birth and growing of inefficient circulation channels, the advancing and position of the water flooding front, and the efficiency of energy, and formulate countermeasures and strategies in time, intervene artificially as frequently as needed, control the growing of inefficient channels, and thus improve the energy efficiency every year. This can play a good role with less money expenditure and better technical effect, and as the result, oilfield development effect can reach the best level.

3. Water Flooding is Suggested as the Main Frame

Water flooding has the advantages of low cost and good technical and economic effects. It is suggested, adhering to the general framework of water flooding, we continuously improve water flooding recovery through innovating the various technological measures to expand the water swept volume and increase the PV of water injected into the swept region. After confirming that the water flooding technology is "eaten dry and squeezed out", then should we consider the implementation of chemical flooding technologies such as polymer flooding.

4. Develop the "Not Stick to the Ground" Thinking Mode and Innovate Transformative Technology

Abandon the "stick to the ground" thinking mode, ponder "out of touch", innovate and investigate transformative technologies, including original innovation and combined innovation. Revolutionary innovation based on "not stick to the ground" thinking mode can lead to earth-shaking and era-epoching technological changes and greatly improve oil recovery. As long as you conceive it, you can do it. Here we give some "not stick to the ground" technical blueprints.

4.1 Micro explosion technology with weakly powerful liquid explosive

Invent a kind of weakly powerful liquid explosive, which can be easily injected into the porous network of reservoir formation, can be detonated relying on local pressurizing or other principles, and explodes at a intensity equivalent to the weak "fizzing" explosion by the small fireworks during the spring festival. This micro explosion technology can not only vary the fluid distribution and activate the remaining oil, but also change the solid structure of the pore and throat network and thus change the flow channels, which is essentially different from the traditional enhanced oil recovery technology.

4.2 Technology of nanometer robot

Drawing the support from modern robot technology, we demonstrate innovative researches, invent nanometer robots capable of continuously advancing through the porous network in reservoir, and thus achieve a substantial increase in oil recovery. This kind of nanometer robot can be equipped with a micro screw propeller at the front end, the propeller being driven by fluid pressure difference, and thus the whole robot can be put in motion to move forward. In the process of movement, it will stimulate the fluid to rotate, start up the remaining oil and drive the fluid flowing forward.

4.3 Role exchanging of oil and water wells

Turn oil well into water well and water well into oil well, completely changing the direction of fluid flow. Maybe, at the beginning of the adjustment, the produced liquid is all water, but after a period of drainage, this kind of role exchanging must be effective. The exchange of roles of oil and water wells is reasonable and tenable from the perspective of the porous flow mechanism. In oilfield running, we need to make a powerful and impetus decision. Firstly, we should start the innovation with the pilot test in small blocks, and gradually sum up the experience and expand the application scale.

5. Management and Supervision are Comparatively Important

Any project, in order to achieve good technical, economic and social results, is inseparable from the management work of "fall onto ground", "put into practice" and "detailing by steps". For example, deal with the reservoir by the technique of grid management. The underground reservoir is divided into tens of thousands, hundreds of thousands, millions, or even tens of millions of grids in the smallest possible size. Engineering management, including static and dynamic data, log and test data, analysis and evaluation, etc., is "talking" to grids, rather than to wells, layers or directions. This is not only a downscaling in size of object, but also, in the more significant meaning, an upgrading in management and supervision work. Such management and supervision work will inevitably strongly support the building of big data, automation, and intellectization, etc., and thus improve the effect and efficiency of oilfield development engineering. During the course of planning, everyone can speak and express their opinions freely. However, once a decision is agreed together, everyone needs to implement the plan without any preconditions and does a good job in putting the plan into specific detail works. In the whole process, it must be assured that supervision is well conducted and thus implementation is finely done. Attention please, doing the best work of management and supervision is of extremely significance for oilfield development engineering.

中文篇

第一章 油田开发核心原理

第一节 能耗原理

油田开发是一个流体消耗能量、克服阻力、产生运动、从地层流动到地面的过程。油田开发全过程都与能量和阻力有关,油田开发效果的好坏也取决于能量和阻力两个方面。

一、开发方式

开发方式,也称为驱动方式,是指地层油依靠何种能量从地层流动到油井井底。开发方式对应渗流过程,指渗流过程中提供能量的方式,包括天然能量和人工补充能量。

1. 天然能量开发

天然能量有以下六种:(1)岩石和流体的膨胀能量;(2)溶解气驱动能量;(3)气顶气能量;(4)天然水驱动能量;(5)重力驱动能量;(6)综合驱动能量。

2. 人工补充能量开发

如果天然能量不足以支持流体克服阻力、从地层流动到油井井底,那么就需要加以人为干预,通过人工注水等途径给地层流体补充能量。人工注水是打专门的注水井,将处理达标的水注入地层,水携带着能量进入地层孔隙网络,接触到油,将能量传递给油,原油获得能量补充,就可以继续克服阻力、向采油井井底流动。

二、能量效率问题

能量是制约油田开发效果的关键要素,涉及能量水平和能量效率两个方面的问题。要想获得好的油田开发效果,一方面要保持较高的能量水平,另一方面要确保较高的能量效率。对于开发时间较长的高含水、特高含水期油田,保持能量水平可能易于实现,想办法提升能量效率才是关键。

由达西定律可知,产液量与能量之间满足以下关系:

$$Q_L = \frac{能量}{阻力} \tag{1-1}$$

能量水平越高,产液量越大。

产油量与能量之间满足以下关系:

$$Q_O = \frac{有效能量}{阻力} = \frac{\alpha \times 能量}{阻力} \tag{1-2}$$

产油量不仅与能量水平有关,而且与能量效率 α 有关。保持能量水平未必能保证产油量,对于高含水、特高含水期油田尤其如此。

如图1-1所示,三个不同渗透率油层构成的多层油藏系统,利用同一注采开发系统进行注水开发。

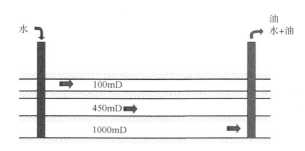

图1-1 多层油藏系统示意图

注入水沿渗透性较好的通道推进较快,通过这些通道首先到达生产井。生产井见水后,这些通道含水饱和度显著增大,渗流阻力减小,所以分配的水量及水的推进速度会明显增大。分配的水量及水的推进速度增大又会造成这些通道含水饱和度继续增大,渗流阻力进一步减小,这又会进一步造成这些通道分配的水量及水的推进速度增大。如此持续,则这些通道最终发展成注入水低效循环通道。

注入水低效循环的形成原因有两方面:一是地质因素的非均质性,如渗透率、厚度、断层发育状况等的非均质性;二是开发因素的非均质性,如注采井距、油水井工作制度等方面的非均质性。

三、提高能量效率的技术途径

提高能量效率的技术途径主要是通过调剖和堵水等工艺措施,封堵低效通道,使其阻力增大,后续注入水转向到剩余油较多的通道,从而提高能量的效率,改善开发效果。

调剖和堵水是应用广泛的工艺措施,技术成熟而且工艺效果也较好,但是,也存在比较严重的缺点。例如,有可能损失一定的可采储量。如图1-2所示,由于孔隙网络是高度复杂的孔隙喉道组合起来的三维体,注入的堵剂封堵低效通道,但是由于低效通道和非低效通道高度复杂、交错衔接,有一部分剩余油富集的孔隙网络就会被堵剂环绕封堵,以后挖潜这一部分剩余油难度加大,甚至会彻底损失。

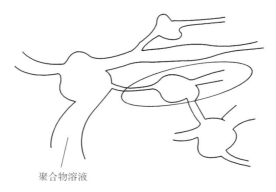

图1-2 调剖和堵水对孔隙网络影响示意图

另外,调剖和堵水措施意味着消耗更多的能量。封堵剩余油少、含水饱和度高、阻力小的低效通道,必然使得整个系统的阻力增大,在产液量不变的情况下,就必然消耗更多的能量。

调剖和堵水措施有显著的提升能量效率的作用,矿场应用的效果非常好。在实践应用过程中,需要重视以上两个问题,持续改进,以便更大程度地提升油田开发水平。

第二节 井网部署

注水开发井网是指注水油田中注水井和采油井的位置及其排布关系,即井网部署。注水开发油田井网部署主要有四种方式:不规则注水开发井网、边缘注水开发井网、规则注水开发井网及顶部底部注水开发井网。

一、不规则注水开发井网

Willhite(1986)指出,地面和地下地形环境、使用定向钻井技术等都会导致注采井分布不均匀。在这些情况下,每口注入井影响的范围大小可能不相同。一些小型油藏采用一次开采方式,部署井数有限,当经济性处于边际时,可能只将少数生产井转为注水井,形成不规则的注水开发井网。受断层影响和孔隙度、渗透率的局部变化制约,也可能会采取不规则注水开发井网。

二、边缘注水开发井网

所谓边缘注水开发井网,就是注入井部署在油藏边部,由边部向内部驱替。典型的边缘注水开发井网如图1-3所示。

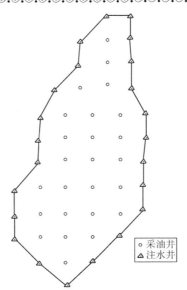

图 1-3 典型的边缘注水开发井网

Craig(1971)在一篇优秀的综述性文献中,指出了边缘注水开发井网的主要特征:

(1)可在最少产水量情况下达到最大产油量。

(2)可将大量产水期推迟到只剩最后一排采油井生产阶段。

(3)和生产井数量相比,注水井数量很少,注入水充满油藏孔隙空间需要较长时间,导致水驱见效缓慢。

(4)为确保边缘注水开发井网取得成效,油藏渗透率应足够高,从而保证注入水以合适速度,从注入井,流过数个排距,到达最后一排采油井。

(5)为使注入井尽可能靠近水驱油前缘,从而使注入水不会绕过任何可驱动原油,可将水淹生产井转为注水井。但是,这需要不断地延伸铺设地面注水管线,增加成本。

(6)边缘注水开发井网的效果难以预测。容易发生驱替流体将油墙推至最内部采油井以里的情况,导致这部分原油难以采出。

(7)因为注入井在边部,注入水需要推进很远的距离,所以保证注入量通常是一个关键问题。

三、规则注水开发井网

由于油田是按照平方英里或者四分之一平方英里来划分区块的,所以井网部署通常也是规则的。

正对式线性井网(图 1-4):注入井和生产井彼此直接正对。这种井网有两个表征参数:a 为同一井排井与井之间的距离,d 为注入井排和生产井排之间的距离。

交错式线性井网(图 1-5):一排注水井,一排采油井,间隔部署。注入井排和生产井排上的井不是直接正对,而是横向位移 $a/2$ 距离,交错开来。

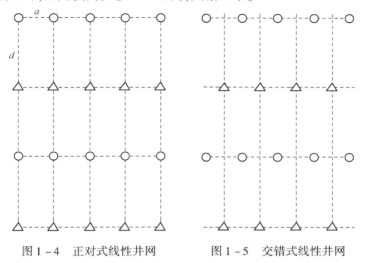

图1-4　正对式线性井网　　图1-5　交错式线性井网

五点法井网(图1-6):是交错式线性井网的特殊情形。所有同类井之间的距离是常数,即 $a=2d$。4口注水井构成一个正方形,生产井位于其中心。

七点法井网(图1-7):注水井位于六边形的6个顶点处,生产井位于其中心。

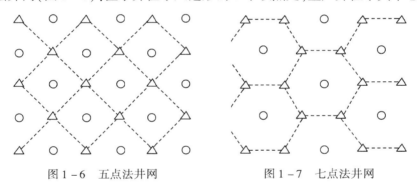

图1-6　五点法井网　　图1-7　七点法井网

九点法井网(图1-8、图1-9):这种注水方式类似于五点法井网,但是在正方形各边的中间位置多布置一口注水井,8口注水井围绕一口生产井。

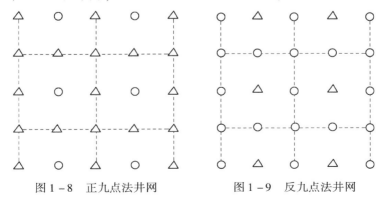

图1-8　正九点法井网　　图1-9　反九点法井网

所谓反式井网是指以注入井为中心来定义井网形式,这是正式和反式井网布置之间的差异。应该注意的是,四点法井网和反七点法井网是相同的(图1-10、图1-11)。

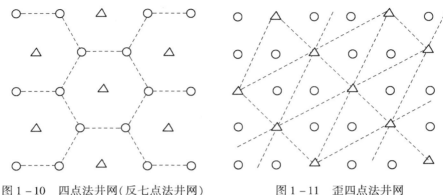

图1-10　四点法井网(反七点法井网)　　　　图1-11　歪四点法井网

四、顶部底部注水开发井网

当构造倾角较大时,顶部底部注水开发井网是较好选择。顾名思义,顶部注水开发井网是将注水井布置在构造顶部,进行油田开发。注入介质为气体时,通常采用顶部注水开发井网。底部注水开发井网是指将注入井布置在构造底部,从构造低部位注入驱替流体。注入介质为水的情况下,一般采用底部注水开发井网,依靠重力分离作用获得更高产量(图1-12)。

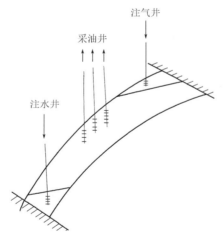

图1-12　倾斜地层井网布置示意图

第三节　气体滞留对水驱采收率的影响

为了研究气体滞留对水驱采收率的影响,进行了大量的实验和矿场研究。早期的研究表明,线性水驱情况下,在注入水之前形成一个油墙,即含油饱和度增大区域。运动的油墙将一部分游离气驱动出来,其余的游离气则滞留下来成为滞留气(图 1 – 13)。多位研究人员通过实验证明,在油藏中建立滞留气含气饱和度 S_{gt},可以提高水驱采收率。

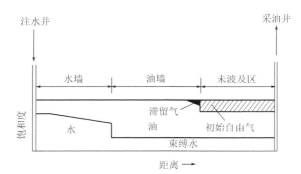

图 1 – 13 水驱期间含水饱和度剖面

关于水驱开发初期存在初始气能够提高水驱采收率这一现象,在理论上还没有成熟的认识。但是,Cole(1969)提出了以下两种不同的理论,或许能对这一现象提供一定的解释。

一、第一种理论

Cole(1969)认为,由于气油界面张力小于气水界面张力,在包含气水油三相的油藏系统中,流体以最小能量关系分布,气体分子将自己封闭在油"毯"中。因为油滴里包裹了一定量的气体,所以油滴的有效尺寸变大了。水驱油过程中,按照流动力学原理,油滴会缩小到某一尺寸。由于油滴内部存在气泡,气泡占据一定体积,所以油藏中残余油实际量变少。

如图 1 – 14 所示,图(a)和图(b)中残余油滴的外观大小相同。但是,在图(b)中,残余油滴的中心不是油,而是气体。因此,图(b)的实际残余油饱和度 S_{or} 比图(a)低。

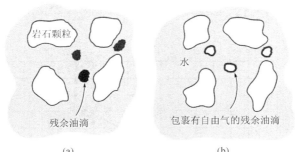

图 1 – 14 游离气饱和度对 S_{or} 的影响(第一种理论)

二、第二种理论

Cole(1969)指出,有一些室内实验报道称,水驱之后再用空气驱替岩心,可以提高采收率。

这些实验所使用的岩心是水湿的。由于毛细管作用,注入水优先进入较小孔隙空间,残余油多位于较大孔隙空间中。水驱之后再接着进行气驱,由于空气为非润湿相,所以气体优先进入较大的孔隙空间。注入气流过这些较大孔隙空间,将一部分水驱后残余油驱替出来。

第二种理论与流体流动观测结果更为接近,原因是气体未必一定存在于油滴中。

如果第二种理论是正确的,那么就可以比较简单地解释游离气的存在使得水湿多孔介质采收率增大这一现象了。由于空气比原油对油藏岩石的润湿性更差,所以,随着不断注入空气,气体将较大孔隙空间的原油驱替出来。

这一现象如图 1-15 所示。在图(a)中,不存在游离气,残余油位于较大孔隙空间内。在图(b)中,存在游离气,游离气挤占了一部分原来石油所占据的空间。图(b)中残余油气饱和度总和近似等于图(a)中残余油饱和度。

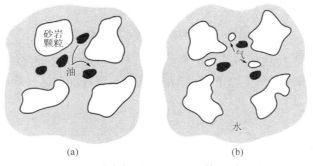

图 1-15 游离气对 S_{or} 的影响(第二种理论)

Craig(1971)建立了两个图版,可以用于计算滞留气导致残余油饱和度的降低值,如图 1-16、图 1-17 所示。其关系如下:

$$S_{gt} = a_1 + a_2 S_{gi} + a_3 S_{gi}^2 + a_4 S_{gi}^3 + \frac{a_5}{S_{gi}}$$

$$\Delta S_{or} = a_1 + a_2 S_{gt} + a_3 S_{gt}^2 + a_5 S_{gt}^3 + \frac{a_5}{S_{gt}}$$

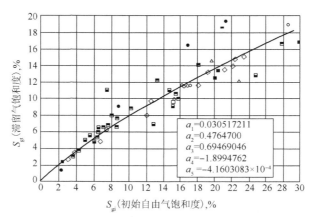

图 1-16 S_{gt} 和 S_{gi} 之间的关系图版

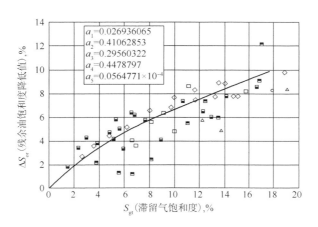

图 1-17　S_{gt} 对水驱采收率的影响

Khelil(1983)认为,如果在水驱开始时存在所谓的"最佳含气饱和度",则可以提高水驱采收率。该最佳含气饱和度由下式给出:

$$(S_g)_{opt} = \frac{0.001867 K^{0.634} B_o^{0.902}}{\left(\dfrac{S_o}{\mu_o}\right)^{0.352} \left(\dfrac{S_{wi}}{\mu_w}\right)^{0.166} \phi^{1.152}} \tag{1-3}$$

式中　$(S_g)_{opt}$——最佳含气饱和度;

S_o, S_{wi}——含油饱和度和初始含水饱和度;

μ_o, μ_w——油和水的黏度,cP;

K——绝对渗透率,mD;

B_o——地层油体积系数;

ϕ——孔隙度。

式(1-3)并不能显性求解,而必须与物质平衡方程(MBE)联合使用。求解 $(S_g)_{opt}$ 的方法是将含气饱和度当作地层压力(或时间)的函数,应用物质平衡方程和式(1-3)同时计算含气饱和度,当两个方程计算的含气饱和度相等时,则认为该含气饱和度即为 $(S_g)_{opt}$。

对于溶解气驱油藏,注入驱替流体通常会导致地层压力回升。如果压力足够高,滞留气体会溶解在原油中,对后续残余油饱和度没有影响。有必要估算将滞留气溶解到原油中所需的压力增量。将滞留气溶解到原油中所需的压力定义为新泡点压力。当压力增加到新泡点压力时,滞留的气体将溶解到油相中,溶解气油比从 R_s 增加到 R_s^{new}。

如图 1-18 所示,新的溶解气油比等于溶解气体积和滞留气体积之和除以地面脱气油体积:

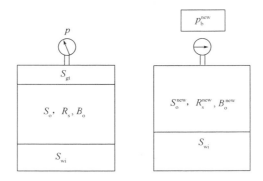

图 1-18 泡点压力变化示意图

$$R_s^{new} = \frac{\dfrac{S_o V_p}{B_o} R_s + \dfrac{S_{gt} V_p}{B_g}}{\dfrac{S_o V_p}{B_o}} \tag{1-4}$$

$$R_s^{new} = R_s + \frac{S_{gt}}{S_o} \frac{B_o}{B_g} \tag{1-5}$$

式中 R_s^{new}——新泡点压力下溶解气油比；

V_p——孔隙体积；

R_s——目前压力 p 下溶解气油比；

B_g——气相体积系数；

B_o——地层油体积系数；

S_{gt}——滞留气饱和度。

在 R_s—p 的关系曲线上，找到与新的溶解气油比（R_s^{new}）对应的压力值，在该压力下，滞留气完全溶解到油相中。

第四节　开发层系

对于多油层油藏，油层多、油层之间差异大、油层段纵向跨度大，有必要细分开发层系，一套开发层系单独用一套井网开发。有的油田开发层系划分多达 7、8 套。

一、划分开发层系的意义

当前各国投入开发的多油层大油田，在大量进行同井分采的同时，基本上采取划分多套开发层系进行开发的方法。

1. 合理划分开发层系，可发挥各类油层的作用

合理划分开发层系，是开发好多油层油田的一项根本措施。所谓划分开发层系，就是

把特征相近的油层组合在一起,采用单独一套开发系统进行开发,并以此为基础进行生产规划、动态研究和调整。

在同一油田内,由于油层在纵向上的沉积环境及其条件不可能完全一致,因而油层特性自然会有差异,所以在开发过程中层间矛盾也就不可避免要出现。若高渗透层和低渗透层合采,则由于低渗透层的流动阻力大,生产能力往往受到限制;低压层和高压层合采,则低压层往往不出油,甚至高压层的油有可能窜入低压层。在水驱油田,高渗透层往往很快水淹,在合采的情况下会使层间矛盾加剧,出现油水层相互干扰,严重影响采收率。图1-19所示为油层倒灌现象。

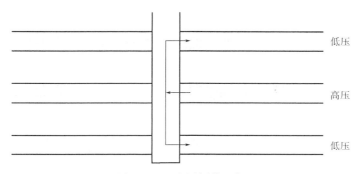

图1-19　油层倒灌现象

2. 划分开发层系是部署井网和规划生产设施的基础

确定了开发层系,就确定了井网套数,因而使得研究和部署井网、注采方式及地面生产设施的规划和建设成为可能。开发区的每一套开发层系,都应独立进行开发设计和调整,对其井网、注采系统、工艺手段等都要独立做出规定,如图1-20所示。

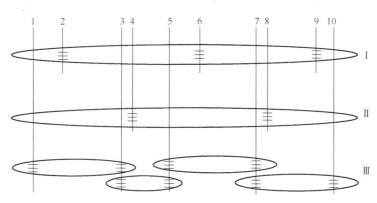

图1-20　开发井网与开发层系匹配示意图

3. 采油工艺技术的发展水平要求进行层系划分

一个多油层油田,其油层数目很多,往往多达几十个,开采井段有时可达数百米。采油工艺的任务在于充分发挥各类油层的作用,使它们吸水均匀、出油均匀,因此,往往采取分层注水、分层采油和分层控制的措施。由于地质条件的复杂性,目前的分层技术还不可能达到很高的水平,因此,就必须划分开发层系,使一个开发层系内部的油层不致过多、井段不致过长。图1-21所示为分采管柱示意图。

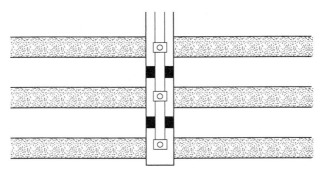

图1-21 分采管柱示意图

4. 油田高速开发要求进行层系划分

用一套井网开发一个多油层油田必须充分发挥各类油层作用,尤其是当主要出油层较多时,为了充分发挥各类油层作用,就必须划分开发层系,这样才能提高采油速度,加快油田的生产,从而缩短开发时间,并提高基本投资的周转率。

二、划分开发层系的原则

总结国内外在开发层系划分方面的经验教训,合理地划分开发层系应考虑的原则是:

(1)把特征相近的油层组合在同一开发层系,以保证各油层对注水开发井网具有共同的适应性,以减少开采过程中的层间矛盾。油层性质相近主要体现在沉积条件相近、渗透率相近、同一层系的油层的分布面积接近、层内非均质程度相近。通常人们以油层组作为组合开发层系的基本单元。有的油田根据大量的研究工作和生产实践,提出以砂岩组来划分和组合开发层系,因为它是一个独立的沉积单元,油层性质相近。

(2)一个独立的开发层系应具有一定的储量,以保证油田满足一定的采油速度,具有较长的稳产时间,并达到较好的经济指标。

(3)各开发层系间必须具有良好的隔层,以便在注水开发的条件下,层系间能够严格地分开,确保层系间不发生串通和干扰。

(4)同一开发层系内油层的构造形态、油水边界、压力系统和原油物性应比较接近。

(5)在分层工艺所能解决的范围内,开发层系不宜划分过细,以便减少建设工作量,提高经济效益。

三、一套开发层系中小层数界限

把层间差异的大小,作为确定组合开发层系小层数的第一因素,换句话来讲就是研究一套开发层系中,能充分发挥各层生产能力的最佳组合。层间干扰是一个复杂的过程,有些因素还在探讨中。

根据某油田 4 口井的单层与多层(自然层)试油资料,分析得到以下 3 组关系曲线:

(1)一套开发层系内,当各小层厚度差别小于 2 倍时,从无量纲有效厚度与无量纲采油指数关系曲线(图 1—22)变化趋势看出,无量纲有效厚度(与第一个层相比的累计倍数)增加到 $h_D<5$ 之前,无量纲采油指数上升较快,当 $h_D>5$ 时,无量纲采油指数上升大大减缓。说明一套开发层系内的有效厚度不大于第一个层的 5 倍时,层间干扰较小,其表达式为:

$$J_D = -0.0152h_D^2 + 0.2709h_D + 0.6455 \tag{1-6}$$

式中 J_D——无量纲采油指数;
h_D——无量纲有效厚度。

式(1—6)的应用条件为 $h_D<9$。

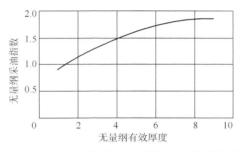

图 1—22 无量纲有效厚度与无量纲采油指数关系曲线

(2)从一套开发层系中的层数与无量纲采油指数关系曲线的变化趋势看出,当一套开发层系内小层数小于 5~7 个时,层间干扰较小,更为合理,如图 1—23 所示。其表达式为:

$$J_D = -0.0117i^2 + 0.2488i + 0.6682 \tag{1-7}$$

式中 J_D——无量纲采油指数;
i——小层层数,个。

式(1—7)的应用条件为 $i<11$ 个。

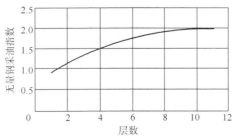

图 1—23 无量纲采油指数与层数关系曲线

(3)一套开发层系内,有效厚度小于12m,若小层间物性接近,最大有效厚度不超过16m,以做到层间干扰较小,油层动用较充分(图1-24),其表达式为:

$$J_D = -0.0019h^2 + 0.0967h + 0.6455 \tag{1-8}$$

式中 h——开发层系有效厚度,m。

式(1-8)的应用条件为 $h < 26$m。

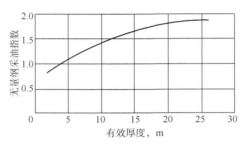

图1-24 无量纲采油指数与有效厚度关系曲线

以上3组曲线是同一个资料统计出来的,从不同侧面分析开发层系组合的定量技术界限,以上3个参数中应该以层数为主,其他均为参考值。

除了以上界限以外,在开发层系划分时,还应该考虑井控储量、层系内油层段纵向跨度、含水饱和度差异等方面的界限。

第五节 经济评价

油田开发经济评价是依据油田开发的方针和原则,在确保获得最高的油田最终采收率的前提下,选择节省投资、经济效益好的油田开发设计方案,从而节约和积累资金。

一、经济评价的任务

油田开发经济评价是分析开发技术方案的经济效益,从而为投资决策提供依据。结合油气田开发工程的特点,油田开发经济评价的主要任务有以下三个方面。

1. 进行工程技术方案的经济评价与可行性研究

油田开发经济评价应配合各级生产管理部门和设计部门做好工程技术方案的经济评价与可行性研究,为提高工程投资项目的综合经济效益提供决策依据。工程技术方案主要包括:(1)新区开发方案;(2)老区调整方案;(3)中外合作开发方案;(4)未开发储量经济评价等。

2. 开展油田开发边际效益分析

为了分析技术方案或技术措施的经济极限,开展油田开发边际效益分析。在实践中,需分析的问题是:(1)极限产能或极限单井日产量;(2)合理井网密度分析;(3)单井极限含水率;(4)热采极限汽油比。

3. 开展油田开发经济动态预测与分析

为了预测油区中长期开发规划的经济效果或分析油田开发经济动态,要做的是:(1)中长期开发规划的经济效果预测;(2)油田开发经济动态分析等。

二、经济评价的依据

开展经济评价,必须以技术指标和投资、成本、费用参数等作为基础,计算关键经济指标。

1. 油田开发技术指标

油田开发技术指标是表征油田生产动态响应特征的参数,包括产油量、含水率、采出程度、地层压力、累积产量等。

计算油田开发指标的方法有物理模拟实验、数值模拟方法、解析方法、经验方法、通用数学方法、测试方法。

进行经济分析或计算所依据的主要技术指标有:(1)油田布井方案,特别是油田的总钻井数、采油井数和注水井数;(2)油田开发阶段的采油速度、采油量、含水上升百分数;(3)各开发阶段的开发年限及总开发年限;(4)不同开发阶段所使用的不同开采方式的井数,即自喷井数及机械采油井数;(5)油田注水或注气方案,不同开发阶段的注水量或注气量;(6)不同开发阶段的采出程度和所预计的采收率;(7)开发过程中的主要工艺技术措施等。

2. 投资、成本和费用参数

(1)投资:需要预测的投资包括勘探投资、开发钻井投资、地面建设投资、系统工程(包括原油储运、输气、油气处理、供电、通信、供排水、道路等)投资、公用工程(包括机修、后勤及辅助企业、矿区民用建设、其他非安装设备、综合利用、环保、计算机等)投资、建设期利息(按复利计算至建设期末)、流动资金等。

(2)成本:油气开采成本项目共划分为14个,包括材料费、工人工资、职工福利、井下作业费、修理费、动力费、其他开采费、测井试井费、燃料费、油气处理费、注水(气)费、轻烃回收费、折旧费、稠油热采费等。

(3)费用。

①管理费用,是指企业行政部门管理和组织经营活动的各项费用,包括公司经费、工会经费、职工教育经费、劳动保险费、待业保险费、董事会费、咨询费、审计费、诉讼费、排污费、绿化费、税金(指房产税、车船使用税、印花税、土地使用税等)、土地损失补偿费、技术转让费、技术开发费、无形资产摊销、开办费摊销、业务费摊销、业务招待费、坏账损失、存货盘亏、

毁损和报废(减盘盈)及其他管理费用。

②矿产资源补偿费,按国家征收规定计算。

③财务费用,是指企业为筹集资金而发生的各项费用,包括企业生产经营期间发生的利息支出(减利息收入)、汇兑净损失、调剂外汇手续费、金融机构手续费及筹资发生的其他财务费用等。

④销售费用,按销售收入的一定比例估算。

三、经济评价指标

经济评价以财务内部收益率、投资回收期、财务净现值等作为主要评价指标。此外,根据项目特点和实际需要,还要计算投资利润率、投资利税率、资金利润率、借款偿还期、流动比率、速动比率等其他指标,以便进行辅助分析。

(1)财务内部收益率。财务内部收益率是指在整个计算期内各年净现金流量现值等于零时的折现率,它反映项目所占用资金的盈利率,是考察项目盈利能力的主要动态评价指标。其表达式为:

$$\sum_{t=1}^{n}(C_1 - C_0)_t(1 + FIRR)^{-t} = 0 \qquad (1-9)$$

式中 C_1——现金流入量;

C_0——现金流出量;

$(C_1 - C_0)_t$——第 t 年的净现金流量;

$FIRR$——财务内部收益率;

n——计算期。

(2)投资回收期。投资回收期是指以项目的净收益抵偿全部投资(固定资产投资、投资方向调节税和流动资金)所需的时间,它是考察项目在财务上投资回收能力的主要静态评价指标。

(3)财务净现值。财务净现值是指项目按行业的基准收益率或设定的折现率,将项目计算期内各年的净现金流量折算到建设期初的现值之和。它是考察项目在计算期内盈利能力的动态指标,表达式为:

$$FNPV = \sum_{t=1}^{n}(C_1 - C_0)_t(1 + i_c)^{-t} \qquad (1-10)$$

式中 $FNPV$——财务净现值;

i_c——年折现率。

第六节　油田开发方案

油田开发方案(oilfield development plan)是指导油田开发全过程的纲领性文件,对油田的开发层系、开发方式、开发井网、开发速度等重大问题进行科学论证,并做出明确规定。完整的油田开发方案包括油藏工程方案、钻井工程方案、采油工程方案和地面工程方案等。油藏工程方案部分主要包括油田概况、油藏地质特征、油藏工程设计和方案实施要求四个方面的内容。

一、油田概况

油田概况主要包括油田的地理位置、气候、水文、交通及周边经济状况,阐述油田的勘探历程和勘探程度,介绍油田开发准备的情况,具体包括发现井、评价井数量及密度,地震工作量及处理技术,地震测线密度及解释成果,取心及分析化验资料,测井及解释成果,地层测试成果,试采及开发试验结果,油田规模及含油层系等内容。

二、油藏地质特征

油藏地质特征包括油藏的构造特征、油层特征、储层特征、流体特征、压力与温度系统、渗流物理特征、天然能量分析、储量计算与评价等。

三、油藏工程设计

油藏工程设计主要包括开发层系设计、开发方式设计、开发井网设计、开发速度设计、开发指标计算、方案评价与优选等内容。油藏工程设计应坚持"少投入、多产出、经济效益最大化"的开发原则。井网部署应坚持"稀井高产"的原则。开发指标是对设计方案在一定开发时期内的产油量、产水量、产气量及地层压力所做的预测性计算结果,目前一般采用油藏数值模拟方法进行计算。方案评价与优选是根据行业标准对各种方案的开发指标进行经济效益计算,然后从中筛选出最佳方案。

四、方案实施要求

根据油田地质特点,对开发方案提出具体的实施要求,一般包括:(1)钻井次序、完井方式、储层保护措施、投产次序、注水方案及程序运行计划要求;(2)开发试验安排及要求;(3)增产措施要求;(4)动态监测要求,包括监测项目和监测内容;(5)HSE等其他要求。

详细的油田开发方案还包括钻井工程方案、采油工程方案和地面工程方案等,可以参阅行业标准。

有了设计周全的油田开发方案,油田开发就有了蓝图,就可以按照设计方案的要求逐步加以实施。

第二章 驱油效率计算方法

第一节 采出程度定义

二次或三次采油方法的采出程度 R_F 由以下广义表达式给出,是三个系数的乘积:

$$R_F = E_D E_A E_V \tag{2-1}$$

则

$$N_p = N_S E_D E_A E_V \tag{2-2}$$

式中 R_F——采出程度;

E_D——驱替效率,又称驱油效率;

E_A——平面波及系数;

E_V——垂向波及系数;

N_S——水驱开始时原始地质储量;

N_p——累积产油量。

决定平面波及系数的主要因素是流体流度、井网类型、平面非均质性和注入流体的总量。垂向波及系数由纵向非均质性、重力分异作用、流体流度和总注入体积等因素决定。

驱油效率是指在任意时刻从波及区采出油量所占比例,可表示为:

$$E_D = \frac{V_S - V_R}{V_S} \tag{2-3}$$

$$E_D = \frac{V_P \frac{S_{oi}}{B_{oi}} - V_P \frac{\overline{S_o}}{B_o}}{V_P \frac{S_{oi}}{B_{oi}}} \tag{2-4}$$

式中 V_S——水驱开始时的油量;

V_R——剩余油量;

S_{oi}——原始含油饱和度;

B_{oi}——原始地层油体积系数;

$\overline{S}_o$——水驱期间任意时刻地层平均含油饱和度。

假设在水驱期间油层孔隙体积恒定,方程简化为:

$$E_D = \frac{S_{oi} - \overline{S}_o}{S_{oi}} \quad (2-5)$$

其中,原始含油饱和度 S_{oi} 由下式给出:

$$S_{oi} = 1 - S_{wi} - S_{gi} \quad (2-6)$$

然而,在已波及区域,气相饱和度被认为是零,因此:

$$\overline{S}_o = 1 - \overline{S}_w \quad (2-7)$$

驱油效率 E_D 也可以用含水饱和度来表示:

$$E_D = \frac{\overline{S}_w - S_{wi} - S_{gi}}{1 - S_{wi} - S_{gi}} \quad (2-8)$$

式中 $\overline{S}_w$——波及区平均含水饱和度;

S_{gi}——初始含气饱和度;

S_{wi}——初始含水饱和度。

如果在水驱开始时不存在自由气体,则方程简化为:

$$E_D = \frac{\overline{S}_w - S_{wi}}{1 - S_{wi}} \quad (2-9)$$

驱油效率 E_D 随水驱进行而不断增大,即随着 $\overline{S}_w$ 增大而增大。由方程可知,当井网平均含油饱和度降低到残余油饱和度 S_{or} 时,也就是当 $\overline{S}_w = 1 - S_{or}$ 时,驱油效率达到最大值。

驱油效率 E_D 随着储层中含水饱和度增大而不断增加。所以,有必要研究一种方法,用来确定波及区平均含水饱和度和累积注水量(或注水时间)的函数关系。

Buckley 和 Leverett(1942)提出了一种方法,称为前缘驱替理论,为建立平均含水饱和度与累积注水量(或注水时间)的函数关系奠定了基础。这个经典理论由两个方程组成:(1)分流方程;(2)前缘推进方程。

接下来讨论前缘驱替理论及其两个方程。

第二节 分 流 方 程

一、分流方程的建立

分流方程的建立归功于 Leverett(1941)。对于两种不混相的流体,如油和水,水的分流

率 f_w(或任何不混相的驱替流体)定义为水流量除以总流量:

$$f_w = \frac{q_w}{q_t} = \frac{q_w}{q_w + q_o} \tag{2-10}$$

式中　f_w——流动液体中水的比例,即水分流率,也称为含水率;

　　　q_t——日产液量;

　　　q_w——日产水量;

　　　q_o——日产油量。

如图 2-1 所示,考虑两种不混相流体(油和水)通过倾斜线性多孔介质的稳态流动。假设一个均质系统,达西方程可以应用于每一相流体:

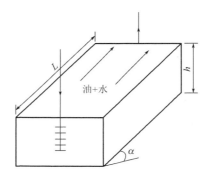

图 2-1　倾斜系统线性驱替

$$q_o = \frac{-K_o A}{\mu_o}\left(\frac{\partial P_o}{\partial x} + g\rho_o \sin\alpha\right) \tag{2-11}$$

$$q_w = \frac{-K_w A}{\mu_w}\left(\frac{\partial P_w}{\partial x} + g\rho_w \sin\alpha\right) \tag{2-12}$$

式中　K_o,K_w——油、水的有效渗透率;

　　　μ_o,μ_w——油、水的黏度;

　　　P_o,P_w——油、水的压力;

　　　ρ_o,ρ_w——油、水的密度;

　　　A——横截面积;

　　　x——距离;

　　　α——倾角,$\sin\alpha$ 上倾方向流动为正,下倾方向流动为负。

下标 o,w 分别代表油和水。

水的分流方程为:

$$f_\text{w} = \frac{1 + \dfrac{K_\text{o}A}{\mu_\text{o}q_\text{t}}\left(\dfrac{\partial P_\text{c}}{\partial x} - g\Delta\rho\sin\alpha\right)}{1 + \dfrac{K_\text{o}}{K_\text{w}}\dfrac{\mu_\text{w}}{\mu_\text{o}}} \tag{2-13}$$

其中 $\Delta\rho = \rho_\text{w} - \rho_\text{o}, P_\text{c} = P_\text{o} - P_\text{w}$

使用油田矿场单位制，式(2-13)可以表示为：

$$f_\text{w} = \frac{1 + \dfrac{0.001127 K_\text{o}A}{\mu_\text{o}q_\text{t}}\left(\dfrac{\partial P_\text{c}}{\partial x} - 0.433\Delta\rho\sin\alpha\right)}{1 + \dfrac{K_\text{o}}{K_\text{w}}\dfrac{\mu_\text{w}}{\mu_\text{o}}} \tag{2-14}$$

$$f_\text{w} = \frac{1 + \dfrac{0.001127\, KK_\text{ro}A}{\mu_\text{o}i_\text{w}}\left(\dfrac{\partial P_\text{c}}{\partial x} - 0.433\Delta\rho\sin\alpha\right)}{1 + \dfrac{K_\text{ro}}{K_\text{rw}}\dfrac{\mu_\text{w}}{\mu_\text{o}}} \tag{2-15}$$

式中　P_c——油水界面张力；

i_w——注水量；

K_ro——油相相对渗透率；

K_rw——水相相对渗透率。

由上述分流方程可见，对于给定的岩石流体系统，除了注水量 i_w、水黏度 μ_w 和流动方向（上倾或下倾注入），方程中其他项由储层特征决定。

方程可用更广义的形式表示，描述任何驱替流体的分流率：

$$f_\text{D} = \frac{1 + \dfrac{0.001127\, KK_\text{rD}A}{\mu_\text{o}i_\text{D}}\left(\dfrac{\partial P_\text{c}}{\partial x} - 0.433\Delta\rho\sin\alpha\right)}{1 + \dfrac{K_\text{ro}}{K_\text{rD}}\dfrac{\mu_\text{D}}{\mu_\text{o}}} \tag{2-16}$$

其中下标 D 表示驱替相流体，$\Delta\rho$ 定义为：

$$\Delta\rho = \rho_\text{D} - \rho_\text{o} \tag{2-17}$$

例如，当驱替流体为不混相的气体时，则：

$$f_\text{g} = \frac{1 + \dfrac{0.001127\, KK_\text{rg}A}{\mu_\text{o}i_\text{g}}\left[\dfrac{\partial P_\text{c}}{\partial x} - 0.433(\rho_\text{g} - \rho_\text{o})\sin\alpha\right]}{1 + \dfrac{K_\text{ro}}{K_\text{rg}}\dfrac{\mu_\text{g}}{\mu_\text{o}}} \tag{2-18}$$

因为毛细管压力梯度一般很小,通常可以忽略,有:

$$f_w = \frac{1 + \dfrac{0.001127 KK_{ro}A}{\mu_o i_w}(0.433\Delta\rho\sin\alpha)}{1 + \dfrac{K_{ro}}{K_{rw}}\dfrac{\mu_w}{\mu_o}} \quad (2-19)$$

$$f_g = \frac{1 + \dfrac{0.001127 KK_{rg}A}{\mu_o i_g}[0.433(\rho_g - \rho_o)\sin\alpha]}{1 + \dfrac{K_{ro}}{K_{rg}}\dfrac{\mu_g}{\mu_o}} \quad (2-20)$$

从含水率的定义,即 $f_w = q_w/(q_w + q_o)$ 可以看出,含水率的极限是 0 和 100%。在束缚水饱和度下,水流量 q_w 为零,因此含水率为 0。在残余油饱和点 S_{or} 处,油流量为零,含水率达到其上限 100%。典型的含水率与含水饱和度关系曲线呈 S 形,如图 2-2 所示。f_w 曲线的界限(0 和 1)可由相对渗透率曲线的端点确定。

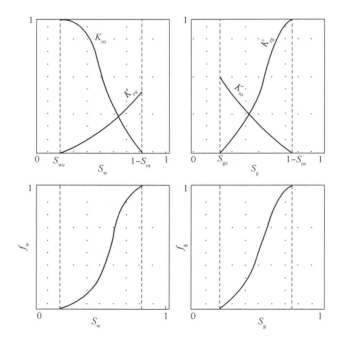

图 2-2 分流率与饱和度关系曲线

二、分流方程的影响因素

1. 水和油黏度的影响

图 2-3 显示了油相黏度对水湿和油湿岩石系统的分流曲线的影响。可以看出,无论系统润湿性如何,油相黏度越大,分流曲线越靠上(增加)。

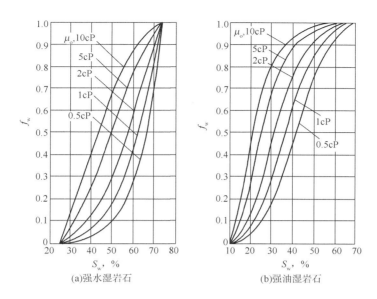

图 2-3　油相黏度对 f_w 的影响

通过分流方程可以清楚地看出，水相黏度对水相分流率(含水率)的影响明显。较高的注入水黏度将导致方程分母的值增大，从而导致 f_w 减小(即向下移动)。

2. 倾角和注入量的影响

为了研究地层倾角 α 和注入量对驱油效率的影响，详细分析分流方程。假设注入量恒定，并且 $(\rho_w - \rho_o)$ 始终为正。为了分别讨论地层倾角和注入量对 f_w 的影响，将分流方程表示为以下简化形式：

$$f_w = \frac{1 - X\dfrac{\sin\alpha}{i_w}}{1 + Y} \tag{2-21}$$

其中，变量 X 和 Y 是不同项的组合，认为是正数，由下式给出：

$$X = \frac{0.001127 \times 0.433 K K_{ro} A (\rho_w - \rho_o)}{\mu_o} \tag{2-22}$$

$$Y = \frac{K_{ro}}{K_{rw}} \frac{\mu_w}{\mu_o} \tag{2-23}$$

对于上倾流动，$\sin\alpha$ 为正。如图 2-4 所示，当水驱油上倾流动(即注入井位于下部)时，效果更好。原因是，这种情况下 $X \sin\alpha / i_w$ 为正值，导致 f_w 下降(向下偏移)。由式(2-21)可知，保持较低的注水量 i_w 是有利的。这是因为，方程右侧的分子 $1 - X \sin\alpha / i_w$ 随着注水量 i_w 降低而减小，所以，随 i_w 降低 f_w 曲线向下移动。

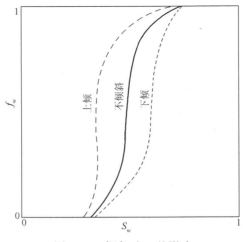

图 2-4　倾角对 f_w 的影响

对于下倾流动,sinα 为负。当水驱油下倾流动(即注入井位于上部)时,$X\sinα/i_w$ 始终为负,因此,方程的分子变为 $1 + X\sinα/i_w$,即：

$$f_w = \frac{1 + X\dfrac{\sinα}{i_w}}{1 + Y} \tag{2-24}$$

这导致 f_w 变大(向上移动)。因此,当注入井位于上部时,提高注水量,可以改善驱替效果。

再分析沿着下倾方向驱替的情形时,将乘积 $X\sinα$ 记为 C,方程可以写为：

$$f_w = \frac{1 + \dfrac{C}{i_w}}{1 + Y} \tag{2-25}$$

式(2-25)表明,如果满足以下条件,则含水率 f_w 可能大于 $1(f_w > 1)$：

$$\frac{C}{i_w} > Y \tag{2-26}$$

只有沿着下倾方向驱替并且注水量较低时,才会出现含水率大于 1 的情况。这种现象称为反向流动,即油相的流动方向与注入水的流动方向相反。当油层有一定倾角、注入井部署在顶部、油井部署在底部时,必须保证注水量足够大以避免油相向着高部位流动。

注意,对于水平油藏,即 $\sinα = 0$,注入量对含水率没有影响。当倾角 $α$ 为零时,方程简化为以下形式：

$$f_w = \frac{1}{1 + \dfrac{K_{ro}}{K_{rw}}\dfrac{\mu_w}{\mu_o}} \tag{2-27}$$

三、含水率与水油比之间的关系

1. 地下含水率(f_w)与地下水油比(WOR_r)之间的关系

$$f_w = \frac{q_w}{q_w + q_o} = \frac{\frac{q_w}{q_o}}{\frac{q_w}{q_o} + 1} \tag{2-28}$$

代入 WOR 得：

$$f_w = \frac{WOR_r}{WOR_r + 1} \tag{2-29}$$

求解 WOR_r 得：

$$WOR_r = \frac{1}{\frac{1}{f_w} - 1} = \frac{f_w}{1 - f_w} \tag{2-30}$$

式中 q_o, q_w ——地下产油量、产水量。

2. 地下含水率(f_w)与地面水油比(WOR_s)之间的关系

$$f_w = \frac{q_w}{q_w + q_o} = \frac{Q_w B_w}{Q_w B_w + Q_o B_o} = \frac{\frac{Q_w}{Q_o} B_w}{\frac{Q_w}{Q_o} B_w + B_o} \tag{2-31}$$

代入 WOR_s 得：

$$f_w = \frac{B_w WOR_s}{B_w WOR_s + B_o} \tag{2-32}$$

求解 WOR_s 得：

$$WOR_s = \frac{B_o}{B_w \left(\frac{1}{f_w} - 1\right)} = \frac{B_o f_w}{B_w (1 - f_w)} \tag{2-33}$$

式中 Q_o, Q_w ——地面产油量、产水量；

B_o, B_w ——地层油、水体积系数。

3. 地下水油比(WOR_r)与地面水油比(WOR_s)之间的关系

从 WOR 的定义知：

$$WOR_r = \frac{q_w}{q_o} = \frac{Q_w B_w}{Q_o B_o} = \frac{\frac{Q_w}{Q_o} B_w}{B_o} \tag{2-34}$$

将 WOR_s 代入上式得：

$$WOR_r = WOR_s \frac{B_w}{B_o} \tag{2-35}$$

$$WOR_s = WOR_r \frac{B_o}{B_w} \tag{2-36}$$

4.地面含水率(f_{ws})与地面油水比(WOR_s)之间的关系

$$f_{ws} = \frac{Q_w}{Q_w + Q_o} = \frac{\dfrac{Q_w}{Q_o}}{\dfrac{Q_w}{Q_o} + 1} \tag{2-37}$$

$$f_{ws} = \frac{WOR_s}{WOR_s + 1} \tag{2-38}$$

5.地面含水率(f_{ws})与地下含水率(f_w)之间的关系

$$f_{ws} = \frac{B_o}{B_w\left(\dfrac{1}{f_w} - 1\right) + B_o} \tag{2-39}$$

如上一节所述,分流方程可以用于确定储层中任意位置的含水率f_w,前提是已知该位置的含水饱和度。然而,问题是如何确定任意位置的含水饱和度。答案是,应用前缘推进方程求取储层中任意位置处的含水饱和度。前缘推进方程可以用来确定水驱期间任意时刻储层中含水饱和度分布。

第三节　前缘推进方程

Buckley和Leverett(1942)提出了非混相两相线性流动基本方程。针对驱替相流体流过任意多孔介质微元,建立物质平衡关系式:

$$\text{流入体积} - \text{流出体积} = \text{体积变化量} \tag{2-40}$$

考虑如图2-5所示的多孔介质微元,微元长度为dx,面积为A,孔隙度为ϕ。

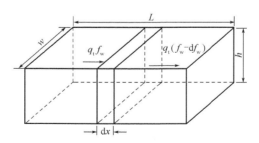

图2-5　水相流过线性微元示意图

在时间微元dt内,流入该微元的总水量为$q_t f_w dt$;流出该微元的流体含水率为($f_w - df_w$),总体积为$q_t(f_w - df_w)dt$;该微元内含水饱和度变化值为dS_w。按照物质平衡关系式,可得:

$$q_t f_w \mathrm{d}t - q_t (f_w - \mathrm{d}f_w) \mathrm{d}t = A\phi \mathrm{d}x \mathrm{d}S_w / 5.615 \tag{2-41}$$

$$q_t \mathrm{d}f_w \mathrm{d}t = A\phi \mathrm{d}x \mathrm{d}S_w / 5.615 \tag{2-42}$$

$$\left(\frac{\mathrm{d}x}{\mathrm{d}t}\right)_{S_w} = (v)_{S_w} = \left(\frac{5.615 q_t}{\phi A}\right)\left(\frac{\mathrm{d}f_w}{\mathrm{d}S_w}\right)_{S_w} \tag{2-43}$$

式中 $(v)_{S_w}$——任意等饱和度面 S_w 的推进速度；

$(\mathrm{d}f_w/\mathrm{d}S_w)_{S_w}$——$f_w$—$S_w$ 关系曲线在 S_w 处的斜率。

上述关系表明，任意等含水饱和度面 S_w 的推进速度与 f_w—S_w 关系曲线在 S_w 处的斜率成正比。注意，对于两相流来说，产液量 q_t 等于注入量 i_w，有：

$$\left(\frac{\mathrm{d}x}{\mathrm{d}t}\right)_{S_w} = (v)_{S_w} = \frac{5.615 i_w}{\phi A}\left(\frac{\mathrm{d}f_w}{\mathrm{d}S_w}\right)_{S_w} \tag{2-44}$$

为了计算在时间 t 内任意等饱和度面推进的距离，必须对方程进行积分：

$$\int_0^x \mathrm{d}x = \frac{5.615 i_w}{\phi A}\left(\frac{\mathrm{d}f_w}{\mathrm{d}S_w}\right)_{S_w} \int_0^t \mathrm{d}t \tag{2-45}$$

$$(x)_{S_w} = \frac{5.615 t i_w}{\phi A}\left(\frac{\mathrm{d}f_w}{\mathrm{d}S_w}\right)_{S_w} \tag{2-46}$$

在恒定的水注入速率下，累积注水量为：

$$W_{\mathrm{inj}} = t i_w \tag{2-47}$$

则可以得到：

$$(x)_{S_w} = \frac{5.615 W_{\mathrm{inj}}}{\phi A}\left(\frac{\mathrm{d}f_w}{\mathrm{d}S_w}\right)_{S_w} \tag{2-48}$$

式中 W_{inj}——累积注水量；

t——时间；

$(x)_{S_w}$——t 时刻任意等饱和度面 S_w 与注入端的距离。

式（2-48）表明，在任意给定累积注水量 W_{inj} 时，等饱和度面 S_w 的位置与 f_w—S_w 曲线在 S_w 处的斜率成正比。在任意给定时刻 t，给定一系列含水饱和度，计算 f_w—S_w 曲线斜率，计算对应的位置，即可绘制该时刻的含水饱和度剖面。

图 2-6 为典型 S 形的 f_w 及其导数的曲线。但是，利用导数曲线计算在任意给定时刻的含水饱和度剖面有一定问题。假设想要计算两个不同的等饱和度面（如图中 A 和 B）在累积注水量达到 W_{inj} 时的位置，应用方程可得：

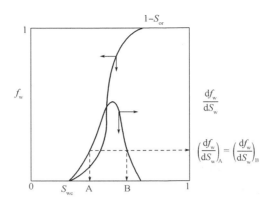

图 2-6 f_w 及其导数曲线

$$(x)_A = \frac{5.615 W_{inj}}{\phi A} \left(\frac{df_w}{dS_w}\right)_A \qquad (2-49)$$

$$(x)_B = \frac{5.615 W_{inj}}{\phi A} \left(\frac{df_w}{dS_w}\right)_B \qquad (2-50)$$

由图 2-6 可以看出,二者的导数相同,即 $(df_w/dS_w)_A = (df_w/dS_w)_B$,这意味着在同一个位置同时存在多个含水饱和度,但从物理意义上来讲这是不可能的。Buckley 和 Leverett (1942)认识到这一点。他们指出,出现这种明显问题的原因是,分流方程中忽略了毛细管压力梯度,这一项表示为:

$$\frac{0.001127 K_o A}{\mu_o i_w} \frac{dP_c}{dx}$$

在绘制分流率曲线时,将上述毛细管压力梯度项考虑进来,可以得到两段曲线,如图 2-7 所示。一段是直线,斜率恒定,从 S_{wc} 到 S_{wf},斜率均为 $(df_w/dS_w)_{S_{wf}}$。另一段是下凹的曲线,从 S_{wf} 到 $(1-S_{or})$,斜率逐渐减小。

Terwilliger 等人(1951)发现,在较低的饱和度范围内(S_{wc} 和 S_{wf} 之间),所有的等饱和度面的推进速度都相等。注意,在这个范围内的所有饱和度对应的斜率都是相同的,因此,由方程可看出,其对应的推进速度也相同:

$$(v)_{S_w < S_{wf}} = \frac{5.615 i_w}{\phi A} \left(\frac{df_w}{dS_w}\right)_{S_{wf}} \qquad (2-51)$$

还可以看出,在这个范围内的所有等饱和度面,在任意给定时刻,其移动的距离也是相等的,如下式所示:

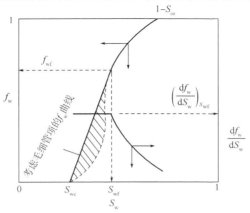

图 2-7　毛细管作用对 f_w 的影响

$$(x)_{S_w < S_{wf}} = \frac{5.615 i_w t}{\phi A} \left(\frac{df_w}{dS_w}\right)_{S_{wf}} \qquad (2-52)$$

结果就是,在 S_{wc} 到 S_{wf} 饱和度范围内,含水饱和度剖面保持恒定形状。Terwilliger 和他的同事,将含水饱和度从 S_{wc} 到 S_{wf} 的区域称为稳定区,其中所有等饱和度面都以相同的速度运动。图 2-8 描述了稳定区的概念。他们还确定了另一个饱和度区间,从 S_{wf} 到 $(1-S_{or})$,其中任意等饱和度面的运动速度都是变化的,称这个区间为非稳定区。

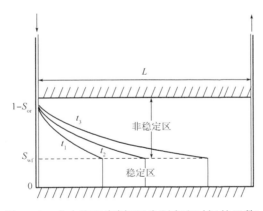

图 2-8　含水饱和度剖面(为距离和时间的函数)

岩心驱油实验表明,实际水驱过程中的含水饱和度剖面与图 2-8 相似,存在一个明显的前缘,或者称作突变前缘。在前缘处,含水饱和度从 S_{wc} 突变到 S_{wf}。在前缘之后,含水饱和度从 S_{wf} 逐渐增加到最大值 $(1-S_{or})$。因此,S_{wf} 称为前缘含水饱和度。

Welge(1952)提出,从 S_{wc}(如果 S_{wc} 与 S_{wi} 不同,则从 S_{wi})画一条与分流率曲线相切的直线,切点对应的含水饱和度值就是前缘含水饱和度 S_{wf}。由切点对应坐标也可确定水驱前缘含水率 f_{wf}。

计算任意给定时刻 t_1 对应的含水饱和度剖面步骤如下:

步骤1：忽略毛管压力项，绘制分流率曲线，即 f_w—S_w 关系曲线。

步骤2：从 S_{wi} 出发，画分流率曲线的切线。

步骤3：确定切点，并读取 S_{wf} 和 f_{wf} 的值。

步骤4：计算切线的斜率 $(df_w/dS_w)_{S_{wf}}$。

步骤5：利用公式计算水驱前缘距注入井的距离：

$$(x)_{S_{wf}} = \frac{5.615 i_w t_1}{\phi A} \left(\frac{df_w}{dS_w}\right)_{S_{wf}} \tag{2-53}$$

步骤6：给定几个大于 S_{wf} 的含水饱和度 S_w 值，在每个给定的含水饱和度处做分流率曲线的切线，确定 $(df_w/dS_w)_{S_w}$，如图2-9所示。

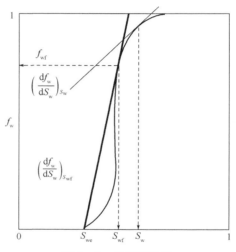

图2-9　分流率曲线

步骤7：应用公式计算从注入井到任意等饱和度面的距离：

$$(x)_{S_w} = \frac{5.615 i_w t_1}{\phi A} \left(\frac{df_w}{dS_w}\right)_{S_w} \tag{2-54}$$

步骤8：使用步骤7得到的结果作图，得到 t_1 时刻的含水饱和度剖面。

步骤9：给定一个新的时间 t_2，重复步骤5到步骤7，得到如图2-8所示的含水饱和度剖面。

在应用图解法确定不同饱和度对应的斜率 $(df_w/dS_w)_{S_w}$ 时，可能会出现一些不规则的结果。认识到相对渗透率 (K_{ro}/K_{rw}) 可以表示为：

$$\frac{K_{ro}}{K_{rw}} = a e^{bS_w} \tag{2-55}$$

那么采用数学方法计算斜率值，结果较好。

注意，式(2-55)中斜率 b 是一个负数。将式(2-55)代入分流率方程，可得：

$$f_w = \cfrac{1}{1 + \cfrac{\mu_w}{\mu_o} a e^{bS_w}} \tag{2-56}$$

对 S_w 求导,可得:

$$\left(\frac{df_w}{dS_w}\right)_{S_w} = \cfrac{-\cfrac{\mu_w}{\mu_o} a b e^{bS_w}}{\left(1 + \cfrac{\mu_w}{\mu_o} a e^{bS_w}\right)^2} \tag{2-57}$$

水驱前缘到达生产井时,即注入水在油井突破时,水驱前缘运动距离刚好等于注水井和油井两者之间的距离。因此,要确定突破时间 t_{BT},只需令 $(x)_{S_{wf}}$ 等于注入井与生产井之间的距离 L,即:

$$L = \frac{5.615 i_w t_{BT}}{\phi A} \left(\frac{df_w}{dS_w}\right)_{S_{wf}} \tag{2-58}$$

注意,孔隙体积为:

$$PV = \frac{\phi A L}{5.615} \tag{2-59}$$

结合以上两式,求解突破时间 t_{BT},得到:

$$t_{BT} = \frac{PV}{i_w} \frac{1}{\left(\dfrac{df_w}{dS_w}\right)_{S_{wf}}} \tag{2-60}$$

假设注水量恒定,则注入水在油井突破时累积注水量为:

$$W_{iBT} = i_w t_{BT} = \frac{PV}{\left(\dfrac{df_w}{dS_w}\right)_{S_{wf}}} \tag{2-61}$$

式中 W_{iBT}——油井见水时对应的累积注水量。

用注入水孔隙体积倍数表示累积注入水量比较方便,即用 W_{inj} 除以储层孔隙体积表示。通常,用 Q_i 表示注入水的总孔隙体积倍数,则在油井见水时:

$$Q_{iBT} = \frac{W_{iBT}}{PV} = \frac{1}{\left(\dfrac{df_w}{dS_w}\right)_{S_{wf}}} \tag{2-62}$$

式中 Q_{iBT}——注水突破时的注入水累积孔隙体积倍数。

为了更好地理解 Buckley 和 Leverett(1942)水驱前缘推进理论的意义,还需要做进一步的讨论。油井见水时累积注水倍数为:

$$W_{iBT} = PV \frac{1}{\left(\dfrac{df_w}{dS_w}\right)_{S_{wf}}} = PV Q_{iBT} \tag{2-63}$$

将分流率曲线的切线外推至 $f_w=1$，对应的含水饱和度为 S_w^*（图 2-10），则切线斜率可由下式计算得到：

$$\left(\frac{\mathrm{d}f_w}{\mathrm{d}S_w}\right)_{S_{wf}} = \frac{1-0}{S_w^* - S_{wi}} \tag{2-64}$$

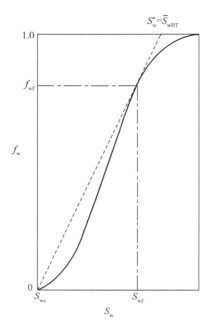

图 2-10　注入水突破时平均含水饱和度

结合以上两个表达式得到：

$$W_{iBT} = PV(S_w^* - S_{wi}) = PVQ_{iBT} \tag{2-65}$$

由式(2-65)可知，含水饱和度 S_w^* 就是注入水突破时油层平均含水饱和度，即：

$$W_{iBT} = PV(\bar{S}_{wBT} - S_{wi}) = PVQ_{iBT} \tag{2-66}$$

式中　$\bar{S}_{wBT}$——注入水前缘突破时平均含水饱和度。

在确定 $\bar{S}_{wBT}$ 时必须考虑以下两点：

一是，在画切线时，若原始含水饱和度 S_{wi} 与束缚水饱和度 S_{wc} 不同，则必须以初始水饱和度 S_{wi} 为起点画切线，如图 2-11 所示。

二是，如果考虑平面波及系数 E_A 和纵向波及系数 E_V，则公式为：

$$W_{iBT} = PV(\bar{S}_{wBT} - S_{wi})E_{ABT}E_{VBT} \tag{2-67}$$

或者

$$W_{iBT} = PVQ_{iBT}E_{ABT}E_{VBT} \tag{2-68}$$

式中　E_{ABT}, E_{VBT}——注入水突破时的平面和纵向波及系数。

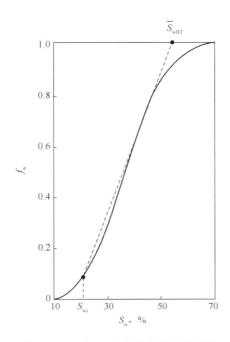

图 2-11 由 S_{wi} 做分流率曲线的切线

注意,在注入水突破之前,波及区域平均含水饱和度将保持 $\bar{S}_{wBT}$ 不变,如图 2-12 所示。在注入水突破时,水驱前缘饱和度 S_{wf} 推进到生产井,含水率由 0 突然增加到 f_{wf}。注入水突破时的 S_{wf} 和 f_{wf} 可以写为 S_{wBT} 和 f_{wBT}。

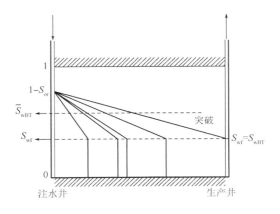

图 2-12 注入水突破前平均含水饱和度

注入水突破后,随着不断注水,生产井处含水饱和度和含水率逐渐增大,如图 2-13 所示。一般把生产井被称为井 2,因此,生产井处含水饱和度和含水率分别记为 S_{w2} 和 f_{w2}。

Welge(1952)指出,注入水突破后,当生产井处含水饱和度达为 S_{w2} 时,可用分流率曲线确定产出液含水率 f_{w2}、平均含水饱和度 $\bar{S}_{w2}$ 和累积注水倍数 Q_i。

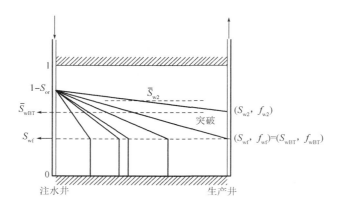

图 2-13 注入水突破后平均含水饱和度

如图 2-14 所示,在任意给定 S_{w2} 大于 S_{wf} 的情况下,绘制分流曲线的切线具有以下特征:

(1)切点对应的分流率值对应生产井含水率 f_{w2}。

(2)切线与 $f_w = 1$ 相交点的饱和度为波及区域平均含水饱和度 $\bar{S}_{w2}$,其数学表达式为:

$$\bar{S}_{w2} = S_{w2} + \frac{1 - f_{w2}}{\left(\dfrac{\mathrm{d}f_w}{\mathrm{d}S_w}\right)_{S_{w2}}} \tag{2-69}$$

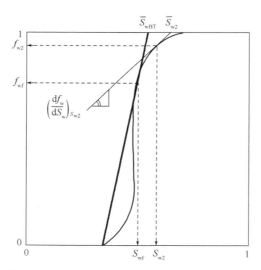

图 2-14 注入水突破后平均含水饱和度的确定

(3)将切线斜率的倒数确定为生产井含水饱和度达到 S_{w2} 时累积注水孔隙体积倍数 Q_i,即:

$$Q_\mathrm{i} = \frac{1}{\left(\dfrac{\mathrm{d}f_\mathrm{w}}{\mathrm{d}S_\mathrm{w}}\right)_{S_\mathrm{w2}}} \tag{2-70}$$

(4)生产井端含水饱和度达到 S_w2 时,累积注水量为:

$$W_\mathrm{inj} = PVQ_\mathrm{i}E_\mathrm{A}E_\mathrm{V} \tag{2-71}$$

或者

$$W_\mathrm{inj} = PV(\overline{S}_\mathrm{w2} - S_\mathrm{wi})E_\mathrm{A}E_\mathrm{V} \tag{2-72}$$

对于恒定的注入量 i_w,累计注入水达到 W_inj 的总时间 t 为:

$$t = \frac{W_\mathrm{inj}}{i_\mathrm{w}} \tag{2-73}$$

第四节 生产动态指标计算

计算采收率的主要目的是在给定注水方案下得到一系列的生产动态曲线。生产动态曲线是开采动态指标与时间的关系图。生产动态指标包括产油量 Q_o、产水量 Q_w、地面水油比 WOR_s、累积产油量 N_p、采出程度 R_F、累积产水量 W_p、累积注水量 W_inj、注水压力 p_inj、注水量 i_w。

一、生产动态计算的指标

生产动态计算一般分为水驱前缘突破前和突破后两部分计算。无论是水驱的哪个阶段,注入水突破前或突破后,累积产油量均由下式表示:

$$N_\mathrm{p} = N_\mathrm{S}E_\mathrm{D}E_\mathrm{A}E_\mathrm{V} \tag{2-74}$$

当 $S_\mathrm{gi} = 0$ 时,驱油效率表达式为:

$$E_\mathrm{D} = \frac{\overline{S}_\mathrm{w} - S_\mathrm{wi}}{1 - S_\mathrm{wi}} \tag{2-75}$$

注入水突破时,E_D 可由突破时平均含水饱和度来计算:

$$E_\mathrm{DBT} = \frac{\overline{S}_\mathrm{wBT} - S_\mathrm{wi}}{1 - S_\mathrm{wi}} \tag{2-76}$$

式中 E_DBT——注入水突破时的驱油效率;

S_wBT——注入水突破时的平均含水饱和度。

水驱前缘突破时,累积产油量为:

$$(N_\mathrm{p})_\mathrm{BT} = N_\mathrm{S}E_\mathrm{DBT}E_\mathrm{ABT}E_\mathrm{VBT} \tag{2-77}$$

式中 $(N_\mathrm{p})_\mathrm{BT}$——注入水突破时的累积产油量。

假设 E_A 和 E_V 均为 100%,式(2-77)可简化为:

$$(N_p)_{BT} = N_S E_{DBT} \tag{2-78}$$

二、生产动态指标的计算步骤

1. 注入水突破前

假设在原始状态下不存在游离气,即 $S_{gi}=0$,那么,注入水突破之前的生产动态计算是比较简单的。累积产油量等于累积注水量,产水量为零($W_p = 0, Q_w = 0$)。

2. 注入水突破后

注入水突破后,给定不同的生产井处含水饱和度值,计算生产动态,具体包括三个阶段:数据准备、注入水突破前生产动态计算和注入水突破后生产动态计算。

1)第一阶段:数据准备

步骤 1:由相对渗透率数据计算相对渗透率比值 K_{ro}/K_{rw},在半对数坐标系中绘制相对渗透率比值与含水饱和度的关系曲线。

步骤 2:找到相对渗透率曲线 K_{ro}/K_{rw} 与 S_w 直线段,确定系数 a 和 b 的值,直线段关系式为:

$$\frac{K_{ro}}{K_{rw}} = a e^{bS_w} \tag{2-79}$$

步骤 3:计算并绘制分流率曲线 f_w,必要时可以考虑重力作用,但忽略毛细管压力梯度。

步骤 4:在 S_{wf} 到 $(1-S_{or})$ 之间选定若干个含水饱和度值,确定每个饱和度对应的斜率 (df_w/dS_w)。分流率曲线斜率与含水饱和度的函数关系如下:

$$\left(\frac{df_w}{dS_w}\right)_{S_w} = \frac{-\dfrac{\mu_w}{\mu_o} a e^{bS_w}}{\left(1 + \dfrac{\mu_w}{\mu_o} a e^{bS_w}\right)^2} \tag{2-80}$$

步骤 5:在笛卡儿坐标系中,利用计算得到的数据点绘制斜率 (df_w/dS_w)—S_w 关系曲线,过这些点画一条平滑曲线。

2)第二阶段:注入水突破前生产动态计算($S_{gi}=0, E_A、E_V=100\%$)

步骤 1:由 S_{wi} 出发做分流率曲线的切线,确定出切点 (S_{wf}, f_{wf}),由切线与 $f_w=1.0$ 的交点,确定出注入水突破时的平均含水饱和度,并确定切线的斜率 $(df_w/dS_w)_{S_{wf}}$。

步骤 2:计算注入水突破时累积注水倍数:

$$Q_{iBT} = \frac{1}{\left(\dfrac{df_w}{dS_w}\right)_{S_{wf}}} = \bar{S}_{wBT} - S_{wi} \tag{2-81}$$

步骤3：假设 E_A 和 E_V 均为100%，计算注入水突破时的累积注水量：

$$W_{iBT} = PV(\bar{S}_{wBT} - S_{wi}) \qquad (2-82)$$

或

$$W_{iBT} = PVQ_{iBT} \qquad (2-83)$$

步骤4：计算注入水突破时的驱油效率：

$$E_{DBT} = \frac{\bar{S}_{wBT} - S_{wi}}{1 - S_{wi}} \qquad (2-84)$$

步骤5：计算注入水突破时累积产油量：

$$(N_p)_{BT} = N_S E_{DBT} \qquad (2-85)$$

步骤6：假设注水量恒定，计算注入水突破时间：

$$t_{BT} = \frac{W_{iBT}}{i_w} \qquad (2-86)$$

步骤7：在小于注入水突破时间内选定若干个时间值，即 $t < t_{BT}$，计算：

$$W_{inj} = i_w t$$

$$Q_o = i_w / B_o$$

$$WOR = 0$$

$$W_p = 0$$

$$N_p = \frac{i_w t}{B_o} = \frac{W_{inj}}{B_o} \qquad (2-87)$$

步骤8：计算注入水突破时地面水油比 WOR_s：

$$WOR_s = \frac{B_o}{B_w \left(\dfrac{1}{f_{wBT}} - 1 \right)} \qquad (2-88)$$

注意，上述 WOR_s 表达式只适用于平面波及效率 E_A 和纵向波及效率 E_V 均为100%时的情况。

3）第三阶段：注入水突破后生产动态计算（$S_{gi} = 0$，E_A、$E_V = 100\%$）

计算水驱前缘突破后生产动态时，需要先给定若干生产井处含水饱和度值，即 S_{w2}，确定每个 S_{w2} 相应的油层内平均含水饱和度。所涉及的具体步骤总结如下：

步骤1：在 S_{wBT} 和 $(1-S_{or})$ 之间选择6~8个不同的 S_{w2}（生产井处 S_w）值，并确定这些 S_{w2} 点对应的 (df_w/dS_w) 值。

步骤 2：对于每个 S_{w2}，计算对应的油层平均含水饱和度和地下含水率：

$$f_{w2} = \frac{1}{1 + \frac{\mu_w}{\mu_o} a e^{bS_{w2}}} \tag{2-89}$$

$$\overline{S}_{w2} = S_{w2} + \frac{1 - f_{w2}}{\left(\dfrac{\mathrm{d}f_w}{\mathrm{d}S_w}\right)_{S_{w2}}} \tag{2-90}$$

步骤 3：计算各个 S_{w2} 对应的驱油效率 E_D：

$$E_D = \frac{\overline{S}_{w2} - S_{wi}}{1 - S_{wi}} \tag{2-91}$$

步骤 4：计算各个 S_{w2} 对应的累积产油量 N_p：

$$N_p = N_S E_D E_A E_V \tag{2-92}$$

假设 E_A 和 E_V 等于 100%，则：

$$N_p = N_S E_D \tag{2-93}$$

步骤 5：确定各个 S_{w2} 对应的累积注水倍数 Q_i：

$$Q_i = \frac{1}{\left(\dfrac{\mathrm{d}f_w}{\mathrm{d}S_w}\right)_{S_{w2}}} \tag{2-94}$$

步骤 6：计算各个 S_{w2} 对应的累积注水量：

$$W_{inj} = PV Q_i \text{ 或 } W_{inj} = PV(\overline{S}_{w2} - S_{wi}) \tag{2-95}$$

注意，假设 E_A 和 E_V 均为 100%。

步骤 7：设注水量 i_w 恒定，计算 W_{inj} 对应的时间：

$$t = \frac{W_{inj}}{i_w} \tag{2-96}$$

步骤 8：根据物质平衡方程计算任意时刻 t 时的累积产水量 W_p。在任意时刻累积注水量等于累积产油量与累积产水量之和：

$$W_{inj} = N_p B_o + W_p B_w \tag{2-97}$$

求解 W_p 得：

$$W_p = \frac{W_{inj} - N_p B_o}{B_w} \tag{2-98}$$

更广义的形式为：

$$W_p = \frac{W_{inj} - (\overline{S}_{w2} - S_{wi}) PV E_A E_V}{B_w} \tag{2-99}$$

需要强调的是，上述所有推导都是建立在一个假设条件的基础上，即从水驱开始到水驱停止，整个过程中，油藏内不存在游离气。

步骤 9：计算每个 f_{w2}（在步骤 2 中确定）对应的地面水油比 WOR_s：

$$WOR_s = \frac{B_o}{B_w \left(\dfrac{1}{f_{w2}} - 1\right)} \quad (2-100)$$

步骤 10：根据以下关系式计算产油量和产水量：

$$i_w = Q_o B_o + Q_w B_w \quad (2-101)$$

将 WOR_s 代入式（2-101）可得：

$$i_w = Q_o B_o + Q_o B_w WOR_s \quad (2-102)$$

求解得：

$$Q_o = \frac{i_w}{B_o + B_w WOR_s} \quad (2-103)$$

$$Q_w = Q_o WOR_s \quad (2-104)$$

步骤 11：将步骤 1 到步骤 10 的计算结果制成表 2-1。

表 2-1　计算结果

S_{w2}	f_{w2}	df_w/dS_w	$\bar{S}_{w2}$	E_D	N_p	Q_i	W_{inj}	t	W_p	WOR_s	Q_o	Q_w
S_{wBT}	f_{wBT}	·	S_{wBT}	E_{DBT}	N_{pBT}	Q_{iBT}	W_{iBT}	t_{BT}	0	·	·	·
·	·	·	·	·	·	·	·	·	·	·	·	·
·	·	·	·	·	·	·	·	·	·	·	·	·
·	·	·	·	·	·	·	·	·	·	·	·	·
$1-S_{or}$	1.0	·	·	·	·	·	·	·	·	100%	0	·

步骤 12：利用计算结果做图，绘制生产动态曲线。

第三章 波及系数及其计算方法

第一节 平面波及系数

平面波及系数 E_A 定义为水淹面积与井网控制面积之比。在注水开发过程中,随着注入量增加,E_A 从零开始稳步增加,直到注入水在采出端突破,之后,E_A 以较慢的速度继续增加。

一、平面波及系数影响因素

平面波及系数大小主要取决于三个因素:(1)流度比 M;(2)注采井网;(3)累积注水量 W_{inj}。

1. 流度比 M

通常,任何流体的流度 λ 定义为有效渗透率与流体黏度之比,即:

$$\lambda_o = \frac{K_o}{\mu_o} = \frac{KK_{ro}}{\mu_o} \tag{3-1}$$

$$\lambda_w = \frac{K_w}{\mu_w} = \frac{KK_{rw}}{\mu_w} \tag{3-2}$$

$$\lambda_g = \frac{K_g}{\mu_g} = \frac{KK_{rg}}{\mu_g} \tag{3-3}$$

式中 $\lambda_o, \lambda_w, \lambda_g$——油、水、气相的流度;

K_o, K_w, K_g——油、水、气相的有效渗透率;

K_{ro}, K_{rw}, K_{rg}——油、水、气相的相对渗透率;

K——绝对渗透率。

式(3-1)至式(3-3)表明,流度 λ 是流体饱和度的强相关函数。流度比 M 定义为驱替液的流度 λ_D 与被驱替液的流度 λ_d 之比,即:

$$M = \frac{\lambda_D}{\lambda_d}$$

对于注水开发油藏：

$$M = \frac{\lambda_w}{\lambda_o}$$

将 λ 表达式代入，得：

$$M = \frac{KK_{rw}}{\mu_w} \frac{\mu_o}{KK_{ro}}$$

简化得到：

$$M = \frac{K_{rw}}{K_{ro}} \frac{\mu_o}{\mu_w} \tag{3-4}$$

Muskat(1946)指出，应用式(3-4)计算 M 时，必须使用以下方法来确定 K_{ro} 和 K_{rw}。

(1)油相相对渗透率 K_{ro}。如图 3-1 所示的水驱油过程，水相推着油相向采出端流动，油相是在未被波及区域，所以应该以初始含水饱和度 S_{wi} 下的 K_{ro} 作为油相相对渗透率。

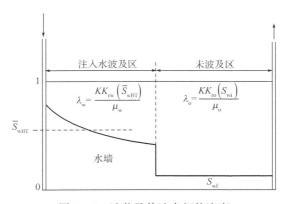

图 3-1　油井见前油水相的流度

(2)水相相对渗透率 K_{rw}。驱替相水相流过之处形成一个水相波及区，其平均含水饱和度为 $\bar{S}_{wBT}$。在注入水突破之前，平均含水饱和度保持不变；注入水突破之后，平均含水饱和度将继续增加(用 $\bar{S}_{w2}$ 表示)。因此，采出端见水前后，属于两个不同阶段，流度比表达式是不一样的。

从开始注水到注入水在采出端突破：

$$M = \frac{K_{rw}(\bar{S}_{wBT})}{K_{ro}(S_{wi})} \frac{\mu_o}{\mu_w} \tag{3-5}$$

式中　$K_{rw}(\bar{S}_{wBT})$——在含水饱和度 $\bar{S}_{wBT}$ 下的水相相对渗透率；

$K_{ro}(S_{wi})$——在含水饱和度 S_{wi} 下的油相相对渗透率。

式(3-5)表明，从开始注水到采出端见水之前，流度比保持恒定。

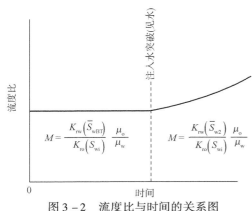

图3-2 流度比与时间的关系图

采出端见水后：

$$M = \frac{K_{rw}(\bar{S}_{w2})}{K_{ro}(S_{wi})} \frac{\mu_o}{\mu_w} \quad (3-6)$$

由式(3-6)可知，注水突破后由于平均含水饱和度$\bar{S}_{w2}$持续增加，水相流度K_w/μ_w会增加，这将导致注水突破后流度比M成比例增加，如图3-2所示。

通常，如果不特别强调，流度比是指注水突破前的流度比。

2. 注采井网

在注水开发方案设计中，通常按照规则的几何形状部署注采井，构建一个对称且相互连通的网状系统。常规的注采井网包括：(1)正对式线性井网；(2)交错式线性井网；(3)五点法井网；(4)七点法井网；(5)九点法井网。

到目前为止，最常用的井网是五点法井网，因此，本章接下来大部分内容将以五点法井网为例进行讨论。

Craig等人(1955)进行了关于流体流度对注水或注气平面波及系数的影响的实验研究。Craig和他的同事利用五点法井网实验室模型(图3-3)，根据不同阶段X射线阴影区面积确定平面波及系数，研究采用了1.43和0.4两个流度比。

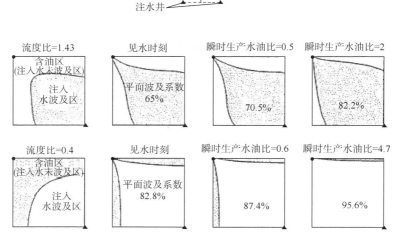

图3-3 水驱过程X射线阴影图

由图3-3可见，在水驱起始阶段，注入水前缘呈现以注水井为中心的圆柱面。随着连续注水，在注入井和生产井之间形成了压力梯度和相应的流线。不同流线的长度不一样，最短流线是注水井和生产井之间的直线。沿这条流线压力梯度最大,注入流体流动比其他流线快。水驱前缘从圆柱面逐渐沿主流线尖突,直至突进到生产井。流度比对平面波及系数的影响很明显。对于流度比为 1.43 的情形,生产井见水时,波及的井区范围比例仅为 65%;对于流度比为 0.4 的情形,生产井见水时,波及的井区范围比例为 82.8%。可以用 E_{ABT} 表示注水突破时的平面波及系数,即注水突破时的水波及面积占比。一般来说,较低流度比情形的平面波及系数较大,而较高流度比的平面波及系数较小。由图 3-3 还可看出,生产井见水后,随着继续注水,平面波及系数继续增加,直至达到 100%。

3. 累积注水量

采出端见水后继续注水,采出程度可大幅提高。尤其是在不利流度比下,更是如此。Craig 等人(1955)研究发现,注水突破后,继续注水可以驱替出大量原油。应当指出,流度比越高,注水突破后的生产阶段就越重要,采油越多。

二、平面波及系数计算方法

计算平面波及系数需分三个阶段:(1)注水突破前;(2)注水突破时;(3)注水突破后。

1. 注水突破前平面波及系数

注水突破前,平面波及系数与注入水量成比例:

$$E_A = \frac{W_{inj}}{PV(\bar{S}_{wBT} - S_{wi})} \tag{3-7}$$

2. 注水突破时平面波及系数

Craig(1955)提出了针对五点法井网的图版,给出了注水突破时平面波及系数 E_{ABT} 与流度比之间的关系曲线,如图 3-4 所示。该图版很近似地模拟了水驱效果,最能代表实际的水驱过程。由图可见,平面波及系数与流度比相关性很强,流度比从 0.15 变化到 10.0,注水突破时平面波及系数从 100% 变为 50%。Willhite(1986)提出了如下数学关公式,该公式可最大程度拟合图 3-4 关系曲线。

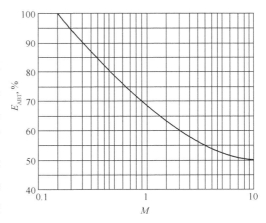

图 3-4 注水突破时平面波及系数

$$E_{ABT} = 0.54602036 + \frac{0.03170817}{M} + \frac{0.30222997}{e^M} - 0.00509693M \tag{3-8}$$

3. 注水突破后平面波及系数

和注水突破后驱油效率 E_D 增加一样,随着继续注水,波及面积增大,平面波及系数不断增大。Dyes 等人(1954)将注水突破后某时刻平面波及系数增加值和该时刻累积注水量 W_{inj} 与注水突破时累积注水量 W_{iBT} 之比建立了关系式：

$$E_A = E_{ABT} + 0.633 \lg \frac{W_{\text{inj}}}{W_{\text{iBT}}} \tag{3-9}$$

或者

$$E_A = E_{ABT} + 0.2749 \ln \frac{W_{\text{inj}}}{W_{\text{iBT}}} \tag{3-10}$$

Dyes 等人还给出了平面波及系数与含水率 f_w 和流度比的倒数 $1/M$ 之间的关系曲线，如图 3-5 所示。

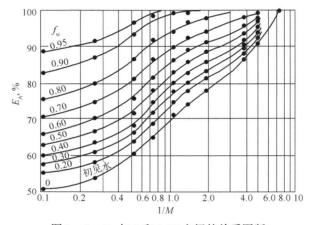

图 3-5 E_A 与 f_w 和 $1/M$ 之间的关系图版

Fassihi(1986)利用非线性回归方法拟合图 3-5 关系曲线，给出了相应的数学关系式：

$$E_A = \frac{1}{1+A} \tag{3-11}$$

其中

$$A = [a_1 \ln(M + a_2) + a_3] f_w + a_4 \ln(M + a_5) + a_6$$

表 3-1 给出了五点法井网、正对式线性井网和交错式线性井网对应的系数。

表 3-1 几种井网的系数

系数	五点法井网	正对式线性井网	交错式线性井网
a_1	-0.2062	-0.3014	-0.2077
a_2	-0.0712	-0.1568	-0.1059
a_3	-0.511	-0.9402	-0.3526
a_4	0.3048	0.3714	0.2608
a_5	0.123	-0.0865	0.2444
a_6	0.4394	0.8805	0.3158

Craig(1971)针对五点注采井网推导出一个公式,给定 E_{ABT} 值,可以计算与 W_{inj}/W_{iBT} 对应的 Q_i/Q_{iBT} 值:

$$\frac{Q_i}{Q_{iBT}} = 1 + E_{ABT} \int_1^x \frac{1}{E_A} dx$$

其中

$$x = \frac{W_{inj}}{W_{iBT}}, Q_i = 1/(df_w/dS_w)_{S_{w2}}, Q_{iBT} = 1/(df_w/dS_w)_{S_{wf}}$$

式中 Q_i——见水后某一时刻累积注水量孔隙体积倍数;

Q_{iBT}——见水时累积注水量孔隙体积倍数。

Craig 计算给出了 Q_i/Q_{iBT} 与 W_{inj}/W_{iBT} 和 E_{ABT} 的关系表,见表 3-2。表中 E_{ABT} 取值范围为 50%~90%,W_{inj}/W_{iBT} 取值范围也很宽。对于任意的 E_{ABT} 和 W_{inj}/W_{iBT} 值,可以根据表 3-2 中数据进行插值,得到对应的 Q_i/Q_{iBT} 值。例如,当 $E_{ABT}=70\%$、$W_{inj}/W_{iBT}=2.00$ 时,可从表 3-2 中确定 Q_i/Q_{iBT} 值为 1.872,即 $Q_i/Q_{iBT}=1.872$。

表 3-2 Q_i/Q_{iBT} 与 E_{ABT} 和 W_{inj}/W_{iBT} 之间的关系数据表

W_{inj}/W_{iBT} \ $E_{ABT}(\%)$	50	51	52	53	54	55	56	57	58	59
1.0	1.000	1.000	1.000	1.000	1.000	1.000	1.000	1.000	1.000	1.000
1.2	1.190	1.191	1.191	1.191	1.191	1.191	1.191	1.191	1.192	1.192
1.4	1.365	1.366	1.366	1.367	1.368	1.368	1.369	1.369	1.370	1.370
1.6	1.529	1.530	1.531	1.532	1.533	1.535	1.536	1.536	1.537	1.538
1.8	1.684	1.686	1.688	1.689	1.691	1.693	1.694	1.696	1.697	1.699
2.0	1.832	1.834	1.837	1.839	1.842	1.844	1.846	1.849	1.851	1.853
2.2	1.974	1.977	1.981	1.984	1.987	1.990	1.993	1.996	1.999	2.001
2.4	2.111	2.115	2.119	2.124	2.127	2.131	2.135	2.139	2.142	2.146
2.6	2.244	2.249	2.254	2.259	2.264	2.268	2.273	2.277	2.282	2.286
2.8	2.373	2.379	2.385	2.391	2.397	2.402	2.407	2.413	2.418	2.422
3.0	2.500	2.507	2.513	2.520	2.526	2.533	2.539	2.545	2.551	2.556
3.2	2.623	2.631	2.639	2.646	2.653	2.660	2.667	2.674	2.681	2.687
3.4	2.744	2.752	2.761	2.770	2.778	2.786	2.793	2.801	2.808	2.816
3.6	2.862	2.872	2.881	2.891	2.900	2.909	2.917	2.926	2.934	2.942
3.8	2.978	2.989	3.000	3.010	3.020	3.030	3.039	3.048	3.057	3.066
4.0	3.093	3.105	3.116	3.127	3.138	3.149	3.159	3.169	3.179	3.189
4.2	3.205	3.218	3.231	3.243	3.254	3.266	3.277	3.288	3.299	3.309
4.4	3.316	3.330	3.343	3.357	3.369	3.382	3.394	3.406	3.417	3.428

续表

W_{inj}/W_{iBT} \ Q_i/Q_{iBT} $E_{ABT}(\%)$	50	51	52	53	54	55	56	57	58	59
4.6	3.426	3.441	3.455	3.469	3.483	3.496	3.509	3.521	3.534	3.546
4.8	3.534	3.550	3.565	3.580	3.594	3.609	3.622	3.636	3.649	
5.0	3.641	3.657	3.674	3.689	3.705	3.720	3.735			
5.2	3.746	3.764	3.781	3.798	3.814	3.830				
5.4	3.851	3.869	3.887	3.905	3.922					
5.6	3.954	3.973	3.993	4.011						
5.8	4.056	4.077	4.097							
6.0	4.157	4.179								
6.2	4.257									
$E_A = 100\%$ 对应的 W_i/W_{iBT}	6.164	5.944	5.732	5.527	5.330	5.139	4.956	4.779	4.608	4.443

W_{inj}/W_{iBT} \ Q_i/Q_{iBT} $E_{ABT}(\%)$	60	61	62	63	64	65	66	67	68	69
1.0	1.000	1.000	1.000	1.000	1.000	1.000	1.000	1.000	1.000	1.000
1.2	1.192	1.192	1.192	1.192	1.192	1.192	1.193	1.193	1.193	1.193
1.4	1.371	1.371	1.371	1.372	1.372	1.373	1.373	1.373	1.374	1.374
1.6	1.539	1.540	1.541	1.542	1.543	1.543	1.544	1.545	1.546	1.546
1.8	1.700	1.702	1.703	1.704	1.706	1.707	1.708	1.709	1.710	1.711
2.0	1.855	1.857	1.859	1.861	1.862	1.864	1.866	1.868	1.869	1.871
2.2	2.004	2.007	2.009	2.012	2.014	2.16	2.019	2.021	2.023	2.025
2.4	2.149	2.152	2.155	2.158	2.161	2.164	2.167	2.170	2.173	2.175
2.6	2.290	2.294	2.298	2.301	2.305	2.308	2.312	2.315	2.319	2.322
2.8	2.427	2.432	2.436	2.441	2.445	2.449	2.453	2.457	2.461	2.465
3.0	2.562	2.567	2.572	2.577	2.582	2.587	2.592	2.597	2.601	2.606
3.2	2.693	2.700	2.705	2.711	2.717	2.723	2.728	2.733	2.738	2.744
3.4	2.823	2.830	2.836	2.843	2.849	2.855	2.862	2.867	2.873	
3.6	2.950	2.957	2.965	2.972	2.979	2.986	2.993			
3.8	3.075	3.083	3.091	3.099	3.107					
4.0	3.198	3.207	3.216	3.225						
4.2	3.319	3.329								
4.4	3.439									
$E_A = 100\%$ 对应的 W_i/W_{iBT}	4.235	4.132	3.984	3.842	3.704	3.572	3.444	3.321	3.203	3.088

续表

W_{inj}/W_{iBT} \ Q_i/Q_{iBT} $E_{ABT}(\%)$	70	71	72	73	74	75	76	77	78	79
1.0	1.000	1.000	1.000	1.000	1.000	1.000	1.000	1.000	1.000	1.000
1.2	1.193	1.193	1.193	1.193	1.193	1.193	1.193	1.194	1.194	1.194
1.4	1.374	1.375	1.375	1.375	1.376	1.376	1.376	1.377	1.377	1.377
1.6	1.547	1.548	1.548	1.549	1.550	1.550	1.551	1.551	1.552	1.552
1.8	1.713	1.714	1.715	1.716	1.717	1.718	1.719	1.720	1.720	1.721
2.0	1.872	1.874	1.875	1.877	1.878	1.880	1.881	1.882	1.884	1.885
2.2	2.027	2.029	2.031	2.033	2.035	2.037	2.039	2.040	2.042	2.044
2.4	2.178	2.180	2.183	2.185	2.188	2.190	2.192	2.195	2.197	
2.6	2.325	2.328	2.331	2.334	2.337	2.340				
2.8	2.469	2.473	2.476	2.480						
3.0	2.610	2.614								
$E_A=100\%$ 对应的 W_i/W_{iBT}	2.978	2.872	2.769	2.670	2.575	2.483	2.394	2.309	2.226	2.147

W_{inj}/W_{iBT} \ Q_i/Q_{iBT} $E_{ABT}(\%)$	80	81	82	83	84	85	86	87	88	89
1.0	1.000	1.000	1.000	1.000	1.000	1.000	1.000	1.000	1.000	1.000
1.2	1.194	1.194	1.194	1.194	1.194	1.194	1.194	1.194	1.194	1.194
1.4	1.377	1.378	1.378	1.378	1.378	1.379	1.379	1.379	1.379	1.379
1.6	1.553	1.553	1.554	1.555	1.555	1.555	1.556	1.556	1.557	1.557
1.8	1.722	1.723	1.724	1.725	1.725	1.726	1.727	1.728		
2.0	1.886	1.887	1.888	1.890						
2.2	2.045									
$E_A=100\%$ 对应的 W_i/W_{iBT}	2.070	1.996	1.925	1.856	1.790	1.726	1.664	1.605	1.547	1.492

W_{inj}/W_{iBT} \ Q_i/Q_{iBT} $E_{ABT}(\%)$	90	91	92	93	94	95	96	97	98	99
1.0	1.000	1.000	1.000	1.000	1.000	1.000	1.000	1.000	1.000	1.000
1.2	1.194	1.195	1.195	1.195	1.195	1.195	1.195	1.195	1.195	1.195
1.4	1.380	1.380	1.380	1.380	1.381					
1.6	1.558									
$E_A=100\%$ 对应的 W_i/W_{iBT}	1.439	1.387	1.338	1.290	1.244	1.199	1.157	1.115	1.075	1.037

Willhite(1986)给出了一个解析式,给定 E_{ABT} 值,即可计算出不同(W_{inj}/W_{iBT})值对应的 (Q_i/Q_{iBT}) 值:

$$\frac{Q_i}{Q_{iBT}} = 1 + a_1 e^{-a_1}[\text{Ei}(a_2) - \text{Ei}(a_1)] \tag{3-12}$$

其中

$$a_1 = 3.65 E_{ABT}, a_2 = a_1 + \ln\frac{W_{inj}}{W_{iBT}}$$

$\text{Ei}(x)$ 是 Ei 函数,可近似为:

$$\text{Ei}(x) = 0.57721557 + \ln x + \sum_{n=1}^{\infty}\frac{x^n}{n(n!)}$$

三、考虑平面波及系数的水驱动态计算

为了在计算水驱动态时把平面波及系数考虑进去,将计算分为三个阶段:(1)初始计算;(2)从开始至注水突破时;(3)注水突破后。上述三个阶段的具体步骤如下。

1. 初始计算($S_{gi}=0, E_V=100\%$)

步骤1:基于相对渗透率数据,计算并在半对数坐标系里绘制相对渗透率比值与含水饱和度关系曲线。用下式描述该直线关系:

$$\frac{K_{ro}}{K_{rw}} = a e^{bS_w}$$

步骤2:计算并绘制 f_w 与 S_w 的关系曲线。

步骤3:从 S_{wi} 出发作分流量曲线的切线,确定:(1)切点(S_{wf}, f_{wf}),即(S_{wBT}, f_{wBT});(2)注水突破时平均含水饱和度 $\bar{S}_{wBT}$;(3)切线的斜率$(df_w/dS_w)_{S_{wf}}$。

步骤4:确定 S_{wi} 和 S_{wBT} 对应的 K_{ro} 和 K_{rw},用 $K_{ro}(S_{wi})$ 和 $K_{rw}(S_{wBT})$ 表示。

步骤5:计算流度比:

$$M = \frac{K_{rw}(\bar{S}_{wBT})}{K_{ro}(S_{wi})}\frac{\mu_o}{\mu_w}$$

步骤6:在 S_{wf} 和 $(1-S_{or})$ 之间选择几个含水饱和度值 S_{w2},计算或者查图确定各饱和度值对应的斜率(df_w/dS_w)。

步骤7:在笛卡儿坐标系里绘制$(df_w/dS_w)_{S_{w2}}$ 与 S_{w2} 的关系曲线。

2. 从开始至注水突破时

假设纵向波及系数 E_V 和初始含气饱和度 S_{gi} 分别为100%和0%,该阶段计算步骤如下:

步骤1:计算注水突破时平面波及系数 E_{ABT}。

步骤2:计算注水突破时累积注水量孔隙体积倍数:

$$Q_{iBT} = \frac{1}{\left(\dfrac{df_w}{dS_w}\right)_{S_{wf}}} = (\bar{S}_{wBT} - S_{wi})$$

步骤 3：计算注水突破时累积注水量 W_{iBT}。

步骤 4：假设注水量为 i_w 恒定，计算注水突破时间 t_{BT}：

$$t_{BT} = \frac{W_{iBT}}{i_w}$$

步骤 5：计算注水突破时的驱油效率 E_{DBT}：

$$E_{DBT} = \frac{\bar{S}_{wBT} - S_{wi}}{1 - S_{wi}}$$

步骤 6：计算注水突破时的累积产油量：

$$(N_p)_{BT} = N_S E_{DBT} E_{ABT}$$

请注意，当 $S_{gi} = 0$ 时，注水突破时的累积产油量等于注水突破时的累积注水量，即：

$$(N_p)_{BT} = \frac{W_{iBT}}{B_O}$$

步骤 7：将 0 到 W_{iBT} 区间划分为任意数量个小区间，对每个小区间计算以下参数：$Q_o = i_w/B_o, Q_w = 0, WOR = 0, N_p = W_{inj}/B_o, W_p = 0, t = W_{inj}/i_w$。

步骤 8：将步骤 1 至 7 数据和结果制成表格（见表 3-3）。

表 3-3 计算结果

W_{inj}	$t = W_{inj}/i_w$	$N_p = W_{inj}/B_o$	$Q_o = i_w/B_o$	WOR_s	$Q_w = Q_o WOR_s$	W_p
0	0	0	0	0	0	0
				0	0	0
				0	0	0
				0	0	0
W_{iBT}	t_{BT}	$(N_p)_{BT}$		WOR_s		0

3. 注水突破后（$S_{gi} = 0, E_V = 100\%$）

Craig 等人（1955）指出，注水突破后，驱替流体继续从已波及区（前缘之后的区域）和新波及区域驱替更多原油。因此，通过将波及区域分成两个不同的区域来计算生产水油比 WOR_s，一是先前波及的区域，二是新波及区域，即驱替流体刚刚波及的区域。

先前波及区域包括含水饱和度大于 S_{wf} 并继续生产油和水的所有区域。随着继续注水，平面波及系数增加，注入水波及更多区域。假设新波及区域仅产油，Craig 等人（1955）提出了一种确定生产水油比 WOR_s 的新方法，该方法的基础是估计总产量为 1bbl（0.159m³）时新波及区域产出的油量，即 ΔN_{pn}，新波及区域生产的油量由下式计算：

$$\Delta N_{\text{pn}} = E\lambda Q_{\text{iBT}} E_{\text{ABT}} \tag{3-13}$$

其中

$$E = \frac{S_{\text{wf}} - S_{\text{wi}}}{E_{\text{ABT}}(\overline{S}_{\text{wBT}} - S_{\text{wi}})}, \lambda = 0.2749 \frac{W_{\text{iBT}}}{W_{\text{inj}}}$$

请注意，参数 E 是常数，而 λ 随着继续注水而减小。Craig 等人（1955）将生产水油比表示为：

$$WOR_{\text{s}} = \frac{f_{\text{w2}}(1 - \Delta N_{\text{pn}})}{1 - [f_{\text{w2}}(1 - \Delta N_{\text{pn}})]} \frac{B_{\text{o}}}{B_{\text{w}}} \tag{3-14}$$

注意，当平面波及系数 E_A 达到 100% 时，新波及面积区域的产油量为零，即 $\Delta N_{\text{pn}} = 0$，可将上述表达式简化为：

$$WOR_{\text{s}} = \frac{f_{\text{w2}}}{1 - f_{\text{w2}}} \frac{B_{\text{o}}}{B_{\text{w}}} = \frac{B_{\text{o}}}{B_{\text{w}}\left(\dfrac{1}{f_{\text{w2}}} - 1\right)}$$

将注水突破后开采动态计算方法总结如下：

步骤 1：给定几个 W_{inj}，$W_{\text{inj}} > W_{\text{iBT}}$。

步骤 2：假设注入量恒定为 i_{w}，计算累积注水量 W_{inj} 所需的时间 t。

步骤 3：对于给定的每一个 W_{inj}，计算对应的 $W_{\text{inj}}/W_{\text{iBT}}$。

步骤 4：对于给定的每一个 W_{inj}，计算平面波及系数 E_A：

$$E_A = E_{\text{ABT}} + 0.6331 \lg \frac{W_{\text{inj}}}{W_{\text{iBT}}} = E_{\text{ABT}} + 0.2749 \ln \frac{W_{\text{inj}}}{W_{\text{iBT}}}$$

步骤 5：计算与每个 $W_{\text{inj}}/W_{\text{iBT}}$ 对应的 $Q_{\text{i}}/Q_{\text{iBT}}$。$Q_{\text{i}}/Q_{\text{iBT}}$ 是 E_{ABT} 和 $W_{\text{inj}}/W_{\text{iBT}}$ 的函数。

步骤 6：将每个 $Q_{\text{i}}/Q_{\text{iBT}}$ 值（在步骤 5 中得到）乘以 Q_{iBT} 来确定累积注水量孔隙体积倍数，即：

$$Q_{\text{i}} = \frac{Q_{\text{i}}}{Q_{\text{iBT}}} Q_{\text{iBT}}$$

步骤 7：根据 Q_{i} 的定义，通过以下方法确定每个 Q_{i} 对应的斜率 $(df_{\text{w}}/dS_{\text{w}})_{S_{\text{w2}}}$：

$$\left(\frac{df_{\text{w}}}{dS_{\text{w}}}\right)_{S_{\text{w2}}} = \frac{1}{Q_{\text{i}}}$$

步骤 8：在 $(df_{\text{w}}/dS_{\text{w}})_{S_{\text{w2}}}$ 和 S_{w2} 关系曲线上，读取每个斜率 $(df_{\text{w}}/dS_{\text{w}})_{S_{\text{w2}}}$ 对应的采出端含水饱和度 S_{w2}（见初始计算阶段，步骤 7）。

步骤 9：计算生产井地下含水率 f_{w2}：

$$f_{\text{w2}} = \frac{1}{1 + \dfrac{\mu_{\text{w}}}{\mu_{\text{o}}} \dfrac{K_{\text{ro}}}{K_{\text{rw}}}}$$

或者：

$$f_{w2} = \frac{1}{1 + \frac{\mu_w}{\mu_o} a e^{bS_{w2}}}$$

步骤10：确定波及区域平均含水饱和度 $\overline{S}_{w2}$：

$$\overline{S}_{w2} = S_{w2} + \frac{1 - f_{w2}}{\left(\frac{df_w}{dS_w}\right)_{S_{w2}}}$$

步骤11：计算每个 S_{w2} 对应的驱油效率 E_D：

$$E_D = \frac{\overline{S}_{w2} - S_{wi}}{1 - S_{wi}}$$

步骤12：计算累积产油量：

$$N_p = N_S E_D E_A E_V$$

对于100%垂向波及系数：

$$N_p = N_S E_D E_A$$

步骤13：计算累积产水量：

$$W_p = \frac{W_{inj} - N_p B_o}{B_w}$$

$$W_p = \frac{W_{inj} - (\overline{S}_{w2} - S_{wi}) PV E_A}{B_w}$$

步骤14：计算与 f_{w2} 对应的地面水油比 WOR_s：

$$WOR_s = \frac{f_{w2}(1 - \Delta N_{pn})}{1 - [f_{w2}(1 - \Delta N_{pn})]} \frac{B_o}{B_w}$$

步骤15：分别计算产油量和产水量：

$$Q_o = \frac{i_w}{B_o + B_w WOR_s}$$

$$Q_w = Q_o WOR_s$$

步骤1至15可以用表3-4形式进行便捷计算。

表3-4 计算结果

W_{inj}	$t = \frac{W_{inj}}{i_w}$	$\frac{W_{inj}}{W_{iBT}}$	E_A	$\frac{Q_i}{Q_{iBT}}$	Q_i	$\left(\frac{df_w}{dS_w}\right)_{S_{w2}}$	S_{w2}	f_{w2}	$\overline{S}_{w2}$	E_D	N_p	W_p	WOR_s	Q_o	Q_w
W_{iBT}	t_{BT}	1.0	E_{ABT}	1.0	Q_{iBT}	—	S_{wBT}	f_{wBT}		E_{DBT}	—	—	—	—	—

注意,迄今为止大家所提出的有关平面波及系数的相关计算公式都是基于理想情形的,都对油藏物性特征设定了苛刻的条件。这些假设条件包括:(1)渗透率均质各向同性;(2)孔隙度均匀分布;(3)油藏无裂缝;(4)封闭边界;(5)饱和度均匀分布;(6)不考虑井网。

为了消除上述假设条件对平面波及系数的影响,通常采用实验结果得到的模型来建立更广义的数学表达式。然而,当不考虑所有的或一部分上述假设条件时,几乎不可能得到广义解。

Landrum 和 Crawford(1960)研究了渗透率方向性对水驱平面波及系数的影响。图 3-6 和图 3-7 给出了线性井网和五点法井网水驱渗透率方向性对平面波及系数的影响。

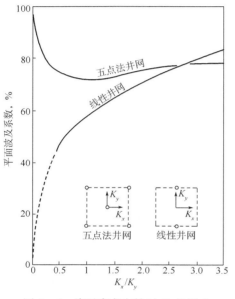

图 3-6　渗透率方向性对 E_A 的影响

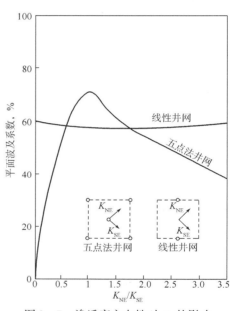

图 3-7　渗透率方向性对 E_A 的影响

四、水驱开发动态计算需要考虑的因素

在开发动态计算中必须考虑三个关键因素:(1)注水量,即流体注入能力;(2)初始含气饱和度;(3)重力分异作用。

1. 注水量

注水量是评价水驱项目时必须考虑的关键经济指标。水驱项目生命周期长短及经济效益好坏受流体注入和产出速度直接影响。估算注水量对于选定合适的注入设备及注入量也很重要。通过小规模先导性水驱试验确定合理注水量是最好的办法。Muskat(1948)和 Deppe(1961)给出了估算规则井网注入能力的经验方法,基于以下假设得到了相关公式:

(1)稳态条件;(2)无初始含气饱和度;(3)流度比为1.0。

注入能力定义为注水量与注采井之间压差的比值,即:

$$I = \frac{i_w}{\Delta P}$$

式中　I——注入能力,bbl/(d·psi);

i_w——注水量,bbl/d;

ΔP——注入压力与生产井井底流动压力之差,psi。

当注入流体与地层油流动性相同(流度比 $M=1$)时,初始注入能力 i_b 表达式为:

$$I_b = \frac{i_b}{\Delta P_b}$$

式中　i_b——初始(基准)注水量,bbl/d;

ΔP_b——注采井初始(基准)压差,psi。

针对完全饱和原油五点井网,即 $S_{gi}=0$,Muskat 提出了如下的注入能力表达式:

$$I_b = \frac{0.003541 h K K_{ro} \Delta P_b}{\mu_o \left(\ln \dfrac{d}{r_w} - 0.619 \right)} \tag{3-15}$$

或者

$$\left(\frac{i}{\Delta P} \right)_b = \frac{0.003541 h K K_{ro}}{\mu_o \left(\ln \dfrac{d}{r_w} - 0.619 \right)} \tag{3-16}$$

式中　i_b——初始(基准)注水量,bbl/d;

h——有效厚度,ft;

K——绝对渗透率,mD;

K_{ro}——含水饱和度为 S_{wi} 时的油相相对渗透率;

ΔP_b——注入井和生产井之间的初始(基准)压差,psi;

d——注入井和生产井之间的距离,ft;

r_w——井筒半径,ft。

为了确定流度比不等于1.0情况下流体注入能力,人们开展了一些研究。从这些研究可以得出以下认识:(1)在有利流度比下,即 $M<1$,随着平面波及系数增加,流体注入能力下降;(2)在不利流度比下,即 $M>1$,随着平面波及系数增加,流体注入能力增加。

Caudle 和 Witte(1959)根据他们的研究结果建立了一个针对五点法井网的流体注入能力与流度比和平面波及系数之间的数学关系式。这个公式仅适用于饱和液体的系统,即 $S_{gi}=0$。引入传导能力比来表示它们之间的关系。传导能力比定义为在水驱任意阶段流体注入能力与初始(基准)注入能力的比值,即:

$$\gamma = \frac{\dfrac{i_w}{\Delta P}}{\left(\dfrac{i}{\Delta P}\right)_b} \tag{3-17}$$

Caudle 和 Witte 以图版形式给出了传导能力比随 E_A 和 M 的变化关系曲线，如图 3-8 所示。注意，如果存在初始含气饱和度，在气体完全溶解或系统完全饱和液体（发生"填充"）之前，该传导能力比图版不适用。

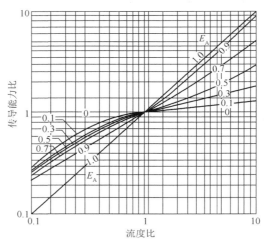

图 3-8　传导能力比图版

式(3-17)适用于如下两种情况。

情况 1：定注入压力变注入量。在恒定注入压力下，$\Delta P_b = \Delta P$，则传导能力比可以写成：

$$\gamma = \frac{i_w}{i_b}$$

或者

$$i_w = \gamma i_b \tag{3-18}$$

情况 2：定注入量变注入压力。当注水量恒定时，即 $i_w = i_b$，传导能力比可表示为：

$$\gamma = \frac{\Delta P_b}{\Delta P}$$

或者

$$\Delta P = \frac{\Delta P_b}{\gamma} \tag{3-19}$$

2. 初始含气饱和度

对于溶解气驱油藏转注水开发的情况，在水驱开始时，油藏中通常存在很高的含气饱和度。有必要先注水一段时间，当累积注水量达到自由气所占孔隙体积时，出口端再开始采油。这个累积注水量称为充填体积。出于经济性考虑,通常会以尽可能大注水量注水,这将导致地层压力上升,足以使得所有滞留气(S_{gt})重新溶解到地层油中。Willhite(1986)指出,常常只需要相对较小幅度的压力上升,就可以重新溶解滞留气体。因而,在水驱开发计算中,通常假设滞留(残余)气饱和度为零。关于五点法井网驱替机理的描述能够说明其他二次采油技术的本质。五点法井网单元由一口生产井和四口注入井构成。四口注入井将原油向内驱替流向中心生产井。如果只有一个五点法井网单元,则注采井数比为 4∶1;但是,对于整个油田而言,包含有数目众多的彼此相邻的五点法井网单元。在这种情况下,注入井与生产井数目的比值接近 1∶1。

溶液气驱油藏水驱起始阶段,油藏中初始含气饱和度高,同时存在油和水,饱和度分别为 S_{oi} 和 S_{wi}。针对油藏中存在初始含气饱和度的情况,Craig、Geffen 和 Morse(1955)将驱替过程分为四个阶段,提出了预测开发动态的方法。该方法以作者名字命名,称为 CGM 方法。他们利用表征四分之一五点法井网的平面模型开展室内实验,基于实验数据建立了 CMG 方法。Craig 等人将水驱过程划分为四个阶段:(1)启动—干扰阶段;(2)干扰—填充阶段;(3)填充—见水阶段;(4)见水—结束阶段,各阶段具体过程如下所述。

1)启动—干扰阶段

溶解气驱油藏开始注水时,井区内通常存在较高含气饱和度,如图 3-9(a)所示。图 3-10 为常规的产量—时间关系曲线。水驱开始前时产油量对应曲线上的 A 点。注水达到一定量时,在注入井周围形成一个高含水饱和度水带。该阶段的特点是水和油均为径向流动。随着继续注水,水带径向驱动油相,形成高含油饱和度油墙。径向流动一直持续到相邻注入井的油墙相遇。相邻注入井的油墙相交的位置称为干扰界,如图 3-11 所示。在这一阶段,生产井周围的情况与水驱刚开始时的情况相似,油井产量 Q_o 没有大的变化,如图 3-10 里面的 B 点所示。Craig、Geffen 和 Morse(1955)给出了径向流扩展这一阶段计算步骤。

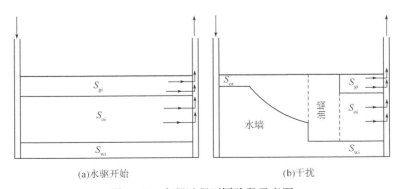

(a)水驱开始 (b)干扰

图 3-9 水驱过程不同阶段示意图

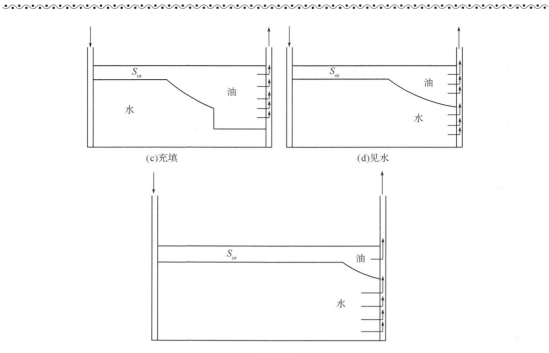

(e) 驱替结束

图 3-9 水驱过程不同阶段示意图(续)

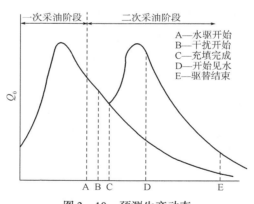

图 3-10 预测生产动态

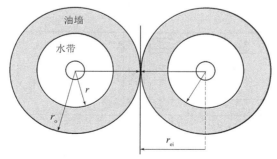

图 3-11 "油墙"相遇

步骤1:根据下式计算截至干扰开始时的累积注水量 W_{ii}:

$$W_{ii} = \frac{\pi h \phi S_{gi} r_{ei}^2}{5.615} \tag{3-20}$$

式中　W_{ii}——截至干扰开始时的累积注水量,bbl;

S_{gi}——初始含气饱和度;

ϕ——孔隙度;

r_{ei}——相邻注入井之间距离的一半,ft。

步骤2:在0和 W_{ii} 之间,给定几个注水量 W_{inj},根据下式计算每个 W_{inj} 相应的注水量:

$$i_w = \frac{0.00707 h K \Delta P}{\frac{\mu_w}{K_{rw}} \ln \frac{r}{r_w} + \frac{\mu_o}{K_{ro}} \ln \frac{r_o}{r}} \tag{3-21}$$

式中　r_o——油墙外径,ft;

r——水带外径,ft;

r_w——井筒半径,ft。

油墙和水带的外半径由以下公式计算:

$$r_o = \sqrt{\frac{5.615 W_{inj}}{\pi h \phi S_{gi}}} \tag{3-22}$$

$$r = r_o \sqrt{\frac{S_{gi}}{\overline{S}_{wBT} - S_{wi}}} \tag{3-23}$$

2) 干扰—填充完成阶段

这一阶段从干扰开始到将原先自由气占据的空间填满为止。填充是水驱产油阶段的开始,如图3-9(c)所示,在图3-10中对应C点。这个阶段流体流动不是严格的径向流动,通常很复杂,难以用数学方法定量描述。因此,只能在填充完成时确定水驱动态。

计算填充完成时开采动态步骤如下:

步骤1:计算截至填充完成时累积注水量 W_{if}:

$$W_{if} = PVS_{gi} \tag{3-24}$$

上述方程表明,当刚开始填充时,与注水量相比,采油量为零或者小到可忽略不计。如果开始填充之前产油量 Q_o 较大,不可忽略,则截至填充完成时累积注水量 W_{if} 必须加上从水驱开始到填充完成这一段时间的总产油量,即:

$$W_{if} = PVS_{gi} + \frac{N_p}{B_o} \tag{3-25}$$

式中　N_p——从水驱开始到填充完成这一段时间的累积产油量,bbl;

B_o——地层油体积系数。

由式(3-25)可以看出,如果考虑填充之前产油量,填充时间将加长,并且需要通过迭代计算确定填充时间。

步骤 2:计算填充阶段平面波及系数:

$$E_A = \frac{W_{inj}}{PV(\bar{S}_{wBT} - S_{wi})}$$

在填充完成时:

$$E_A = \frac{W_{if}}{PV(\bar{S}_{wBT} - S_{wi})}$$

步骤 3:基于填充完成时的流度比和平面波及系数,确定传导能力比 γ。需要注意的是,只有在整个井网系统完全充满液体时,即填充完成时,才可以计算确定传导能力比。

步骤 4:对于恒定压差情况,初始(基准)注水量 i_b 为:

$$i_b = \frac{0.003541 h K K_{ro} \Delta P}{\mu_o \left(\ln \frac{d}{r_w} - 0.619 \right)}$$

步骤 5:计算填充完成时注水量 i_{wf}:

$$i_{wf} = \gamma i_b$$

上式仅在井网系统充满液体时有效,即在填充完成及之后的过程有效。

步骤 6:计算从干扰开始到填充完成的时间长度:

$$\Delta t = \frac{W_{if} - W_{ii}}{\dfrac{i_{wi} + i_{wf}}{2}}$$

上式表明,在干扰开始之后即填充阶段。

3)填充完成—注水突破阶段

如图 3-10 中 C 点所示,填充完成标志着:(1)注水井区内无剩余自由气;(2)油墙前缘到达生产井;(3)采出端见到注水效果;(4)产油量 Q_o 等于注水量 i_w。

在这一阶段,由于波及区内不存在自由气,采油量基本等于注水量。随着不断注水,水带前缘最终到达生产井,标志着注入水突破,如图 3-9(d)所示。注入水在采出端突破后,产水量迅速上升。

这一阶段水驱动态计算步骤如下:

步骤 1:计算注水突破时累积注入水量:

$$W_{iBT} = PV(\bar{S}_{iBT} - S_{wi})E_{ABT} = PV Q_{iBT} E_{ABT}$$

步骤 2:在 W_{if} 和 W_{iBT} 之间,给定几个累积注水量值 W_{inj},计算每个 W_{inj} 值对应的平面波及系数:

$$E_A = \frac{W_{inj}}{PV(\bar{S}_{wBT} - S_{wi})}$$

步骤 3：对每个 W_{inj} 值，确定对应的传导能力比 γ。

步骤 4：计算每个 W_{inj} 对应的注水量：

$$i_w = \gamma i_b$$

步骤 5：计算产油量 Q_o：

$$Q_o = \frac{i_w}{B_o}$$

步骤 6：计算累积产油量 N_p：

$$N_p = \frac{W_{inj} - W_{if}}{B_o}$$

4）注水突破—水驱结束阶段

注水突破后，水油比迅速增加，产油量明显下降，如图 3-10 中 D 点所示。随着注水继续，波及面积继续增加，新波及区域产油，而先前波及区域既产油也产水。

依据已波及区产油量产水量及新波及区产油量计算 WOR。假定新波及区原油是由位于稳定区后面、含水饱和度为 S_{wf} 的等饱和度面驱动的。这个阶段计算如下：

步骤 1：给定若干个与表中给出值相对应的 W_{inj}/W_{iBT} 值，即，1、1.2、1.4 等。

步骤 2：计算每个 W_{inj}/W_{iBT} 值对应的累积注水量：

$$W_{inj} = \frac{W_{inj}}{W_{iBT}} W_{iBT}$$

步骤 3：计算每个 W_{inj}/W_{iBT} 值对应的平面波及系数：

$$E_A = E_{ABT} + 0.6331 \lg \frac{W_{inj}}{W_{iBT}}$$

步骤 4：计算与每个 W_{inj}/W_{iBT} 值相对应的 Q_i/Q_{iBT} 值。

步骤 5：将每个 Q_i/Q_{iBT} 值乘以 Q_{iBT}，确定注入水孔隙体积倍数：

$$Q_i = \frac{Q_i}{Q_{iBT}} Q_{iBT}$$

步骤 6：根据 Q_i 的定义，确定每个 Q_i 值对应的斜率 $(df_w/dS_w)_{S_{w2}}$：

$$\left(\frac{df_w}{dS_w}\right)_{S_{w2}} = \frac{1}{Q_i}$$

步骤 7：从 $(df_w/dS_w)_{S_{w2}}$ 与 S_{w2} 关系曲线读取对应于每个斜率的 S_{w2} 值，即生产井含水饱和度。

步骤 8：计算每个生产井 S_{w2} 对应的地层含水率 f_{w2}：

$$f_{w2} = \frac{1}{1 + \frac{\mu_w}{\mu_o}\frac{K_{ro}}{K_{rw}}}$$

或者：

$$f_{w2} = \frac{1}{1 + \frac{\mu_w}{\mu_o}ae^{bS_{w2}}}$$

步骤 9：确定波及区平均含水饱和度：

$$\overline{S}_{w2} = S_{w2} + \frac{1 - f_{w2}}{\left(\frac{df_w}{dS_w}\right)_{S_{w2}}}$$

步骤 10：计算 f_{w2} 对应的地面水油比：

$$WOR_s = \frac{f_{w2}(1 - \Delta N_{pn})}{1 - f_{w2}(1 - \Delta N_{pn})}\frac{B_o}{B_w}$$

步骤 11：Craig、Geffen 和 Morse（1955）指出，该阶段计算累积产油量时，必须考虑流入未受波及区而损失的油量。为了考虑这部分损失，提出了以下表达式：

$$N_p = N_S E_D E_A - \frac{PV(1 - E_A)S_{gi}}{B_o}$$

其中

$$E_D = \frac{\overline{S}_w - S_{wi} - S_{gi}}{1 - S_{wi} - S_{gi}}$$

步骤 12：根据产出水 = 注入水 - 产出油 - 填充体积计算累积产水量，即：

$$W_p = \frac{W_{inj} - N_p B_o - PVS_{gi}}{B_w}$$

步骤 13：计算 $K_{rw}(\overline{S}_{w2})$，并确定注水突破后的流度比 M：

$$M = \frac{K_{rw}(\overline{S}_{w2})}{K_{ro}(S_{wi})}\frac{\mu_o}{\mu_w}$$

步骤 14：计算传导能力比。

步骤 15：确定注水量：

$$i_w = \gamma i_{base}$$

步骤 16：分别计算产油量和产水量：

$$Q_o = \frac{i_w}{B_o + B_w WOR_s}$$

$$Q_w = Q_o WOR_s$$

3. 重力分异作用

对于原油黏度高、油层厚度大、地层倾斜的水驱油藏，注入水易于沿油层底部舌进。相似地,气驱油情况下,由于气体和原油重力差异,会导致气体超越原油一直向上运动,直至遇到页岩或者低渗透率的纵向遮挡才会停止。在线性水驱实验研究发现,当油水两相的黏度相等时,流体界面保持水平而且和流体速度无关。如果油和水黏度不一样,原始水平界面将变得倾斜。

在倾斜油层研究中,Dake(1978)建立了重力分异模型,使用该模型能够计算稳定驱替所需的临界注水量 i_c。稳定驱替的条件是,整个驱替过程中流体界面和流动方向之间的角度保持恒定,如图3-12所示。Dake引入了两个参数,一个是无量纲重力数 G,一个是末点流度比 M^*,用这两个参数可以定义驱替过程的稳定性。

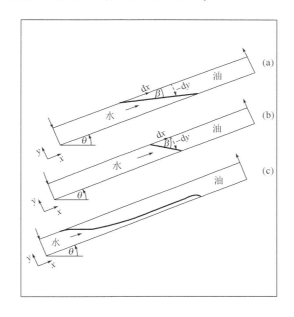

图3-12 考虑重力分异作用的稳定驱替和不稳定驱替
(a)稳定: $G > m - 1, M^* > 1, \beta < \theta$;(b)稳定: $G > M^* - 1, M^* < 1, \beta > \theta$;(c)不稳定: $G < M^* - 1$

这两个参数定义如下:
(1)无量纲重力数。无量纲重力数 G 由下式给出:

$$G = \frac{7.853 \times 10^{-6} K K_{rw} A (\rho_w - \rho_o) \sin\theta}{i_w \mu_w} \tag{3-26}$$

式中 K——绝对渗透率,mD;

K_{rw}——S_{or} 对应的水相相对渗透率;

A——横截面积,ft²;

ρ_w——水的密度,lb/ft³;

θ——油层倾角,(°)。

(2)末点流度比。末点流度比 M^* 由下式定义：

$$M^* = \frac{K_{rw}(S_{or})\mu_o}{K_{ro}(S_{wi})\mu_o} \tag{3-27}$$

Dake 使用上述两个参数定义了以下稳定性标准：

(1)当 $M^* > 1$ 时。如果 $G > M^* - 1$，则运移是稳定的，在这种情况下，流体界面角 $\beta < \theta$；如果 $G < M^* - 1$，则运移不稳定。

(2)当 $M^* = 1$ 时。这是一个非常有利的条件，因为在此情况下水不会绕过油。此时驱替过程是无条件稳定的，此时 $\beta = \theta$。

(3)当 $M^* < 1$ 时。当末点流度比 M^* 小于 1 时，为无条件稳定驱替，此时 $\beta > \theta$。

Dake 还定义了临界注水量 i_c：

$$i_c = \frac{7.853 \times 10^{-6} K K_{rw} A(\rho_w - \rho_o)\sin\theta}{\mu_w (M^* - 1)} \tag{3-28}$$

第二节　纵向波及系数

纵向波及系数 E_V 定义为注入流体垂向上波及油层厚度的比例。纵向波及系数主要取决于流度比和注入流体总体积。由于渗透率非均质性，注入流体在油藏中的流动前缘并不是规则整齐的。在渗透率较高的地方，注入水的流动速度较大；在渗透率较低的地方，注入水的流动速度较小。

一、渗透率非均质性与表征

在进行水驱设计时，最大的不确定性在于对油藏渗透率分布的定量认识程度。迄今为止，渗透率分布被认为是影响纵向波及系数的最重要因素。

为了计算纵向波及系数，工程师必须能够解决以下三个问题：(1)使用数学方法描述和确定渗透率在空间上的分布；(2)确定充分模拟流体流动动态所需最少油层数目；(3)求取各层的平均性质(称为分区问题)。

下面详细讨论上述三个问题。

1. 纵向非均质性

油藏工程师面对的首要问题之一就是处理和使用由岩心及测井分析得到的大量数据。虽然一个油藏中孔隙度和束缚水饱和度在平面与纵向上是变化的，但是影响水驱动态的最重要岩石性质是渗透率。由于不同油层渗透差别通常超过一个数量级，所以描述渗透率涉及一些特殊问题。

Dykstra 和 Parsons(1950)提出了渗透率变异系数这一术语,用以描述储层非均质程度。其取值范围为从 0 到 1。对于完全均质油藏,渗透率变异系数为 0;对于完全非均质油藏,其值为 1。渗透率变异系数 V 表示为:

$$V = \frac{K_{50} - K_{84.1}}{K_{50}}$$

式中　K_{50}——累积百分比 50% 对应的渗透率值;

　　　$K_{84.1}$——累积百分比 84.1% 对应的渗透率值。

为进一步说明 Dykstra 和 Parsons 渗透率变异系数的用途,Craig(1971)做了一个研究。假定有一个油藏,部署了 10 口井(井 A 到井 J),每口井详细渗透率数据见表 3-5。每口井有 10 个渗透率值,每个值对应 1 英尺油层厚度。

表 3-5　假想的 10 层油藏(单位:mD)

深度(ft)	假想油藏岩心分析资料(岩心来自 10 口井,井 A 到井 J,每个渗透率代表 1ft 厚度)									
	A	B	C	D	E	F	G	H	I	J
6791	2.9	7.4	30.4	3.8	8.6	14.5	39.9	2.3	12.0	29.0
6792	11.3	1.7	17.6	24.6	5.5	5.3	4.8	3.0	0.6	99.0
6793	2.1	21.2	4.4	2.4	5.0	1.0	3.9	8.4	8.9	7.6
6794	167.0	1.2	2.6	22.0	11.7	6.7	74.0	25.5	1.5	5.9
6795	3.6	920.0	37.0	10.4	16.5	11.0	120.0	4.1	3.5	33.5
6796	19.5	26.6	7.8	32.0	10.7	10.0	19.0	12.4	3.3	6.5
6797	6.9	3.2	13.1	41.8	9.4	12.9	55.2	2.0	5.2	2.7
6798	50.4	35.2	0.8	18.4	20.1	27.8	22.7	47.4	4.3	66.0
6799	16.0	71.5	1.8	14.0	84.0	15.0	6.0	6.3	44.5	5.7
6800	23.5	13.5	1.5	17.0	9.8	8.1	15.4	4.6	9.1	60.0

Craig(1971)将所有 100 个渗透率按照由大到小顺序排列,做出表征渗透率分布的对数—概率曲线,如图 3-13 所示。渗透率分布表明,该假想油藏的渗透率变异系数为 70%:

$$V = \frac{K_{50} - K_{84.1}}{K_{50}} = \frac{10 - 3}{10} = 0.7$$

2. 层数最小值

Craig(1971)基于计算机开展研究,提出了预测水驱油藏动态所需分层数最小值的确定方法。他模拟了一个 100 层油藏五点法水驱的动态。油藏的渗透率变异系数在 0.1~0.8 之间取值。拟合 100 层油藏模型结果所需的最小层数是流度比 M 和渗透率变异系数 V 的函数。表 3-6 到表 3-8 给出了这些模拟结果,为五点法井网水驱最少层数的确定提供了依据。

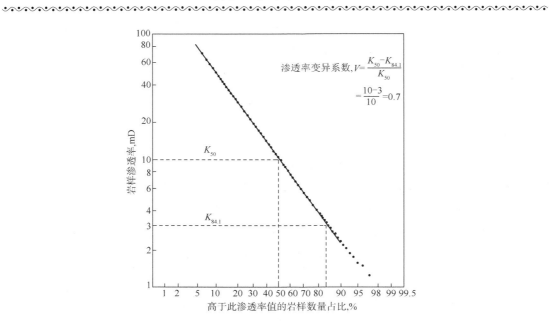

图 3-13 假想油藏渗透率变异系数确定方法

表 3-6 WOR>2.5 情况下最少层数

流度比	渗透率变异系数							
	0.1	0.2	0.3	0.4	0.5	0.6	0.7	0.8
0.05	1	1	2	4	10	20	20	20
0.1	1	1	2	4	10	20	100	100
0.2	1	1	2	4	10	20	100	100
0.5	1	2	2	4	10	20	100	100
1.0	1	3	3	4	10	20	100	100
2.0	2	4	4	10	20	50	100	100
5.0	2	5	10	20	50	100	100	100

表 3-7 WOR>5 情况下最少层数

流度比	渗透率变异系数							
	0.1	0.2	0.3	0.4	0.5	0.6	0.7	0.8
0.05	1	1	2	4	5	10	10	20
0.1	1	1	2	4	10	10	10	100
0.2	1	1	2	4	10	10	20	100
0.5	1	2	2	4	10	10	20	100
1.0	1	2	3	4	10	10	20	100
2.0	2	3	4	5	10	10	50	100
5.0	2	4	5	10	20	100	100	100

表 3-8 WOR > 10 情况下最少层数

流度比	渗透率变异系数							
	0.1	0.2	0.3	0.4	0.5	0.6	0.7	0.8
0.05	1	1	1	2	4	5	10	20
0.1	1	1	1	2	5	5	10	20
0.2	1	1	2	3	5	5	10	20
0.5	1	1	2	3	5	5	10	20
1.0	1	1	2	3	5	10	10	50
2.0	1	2	3	4	10	10	20	100
5.0	1	3	4	5	10	100	100	100

3. 分区问题

在水驱计算中,常常需要把油藏划分成若干个厚度相等孔渗不等的小层。有两种常用方法,可以计算各小层平均渗透率,一是位置法,二是渗透率排序法。

1）位置法

位置法根据在垂向上的相对位置表示各小层。这种方法假定,注入流体从注入端流向采出端过程中,流体保持在同一小层内流动,不发生层间窜流。Miller 和 Lents（1966）在预测 Bodcaw 油藏循环注入项目的开发动态时,成功地论述了这一方法。某个小层的平均渗透率可以采用几何平均算法求得:

$$K_{avg} = \exp\left(\frac{\sum_{i=1}^{n} h_i \ln K_i}{\sum_{i=1}^{n} h_i}\right)$$

如果各点位的厚度都相等,则:

$$K_{avg} = (K_1 K_2 K_3 \cdots K_n)^{1/n}$$

2）渗透率排序法

渗透率排序法以 Dykstra 和 Parsons（1950）方法为基础,将岩心分析渗透率按由大到小顺序排列,并绘制如图 3-13 所示的曲线。将概率轴划分成均匀刻度,每一段刻度区间对应一个小层,该刻度区间中间点对应的渗透率就是该小层的渗透率。

可以采取相似方法确定每个小层的孔隙度。将岩心分析得到的孔隙度数据按照由大到小顺序排序,在笛卡儿—概率坐标系中绘制孔隙度和厚度累积占比之间的关系曲线。将概率轴划分成均匀刻度,每一段刻度区间对应一个小层,该刻度区间中间点对应的孔隙度就是该小层的孔隙度。

在研究纵向波及系数时,渗透率排序法是使用最广泛的方法。

二、纵向波及系数的计算

计算纵向波及系数 E_V 主要使用两种经典方法：Stiles 法和 Dykstra – Parsons 法。这两种方法假定油藏是理想的层状的系统。采用渗透率排序法确定各小层的渗透率，按照渗透率由大到小确定小层排列顺序。

这两种方法共同的假设条件包括：层间无窜流；非混相驱替；线性流动；注入水在各小层渗流距离与该小层渗透率成比例；活塞式驱替。

Stiles 法和 Dykstra – Parsons 法的基本思想都是计算在各小层依次见水过程中所有小层水驱油前缘位置。如果定义各小层地层系数为渗透率与厚度的乘积，即 Kh，那么，就可以计算各个小层的水流量和油流量，从而得到生产水油比。

1. Stiles 法

Stiles（1949）建立了一种预测水驱动态的方法，考虑了渗透率变异系数的影响。Stiles 假定，在一个多层油藏水驱过程中，渗透率最大的小层最先见水，各小层按照渗透率由高到低依次见水。假设油藏分成 n 个小层，这些小层按照渗透率大小降序排列，那么，当第 i 小层见水时，从第 1 层到第 i 层都已水淹，而其他小层还未见水。

基于上述概念，Stiles 提出，纵向波及系数可通过下式计算：

$$E_V = \frac{K_i \sum_{j=1}^{i} h_j + \sum_{j=i+1}^{n} (Kh)_j}{K_i h_t} \tag{3-29}$$

式中 i——见水层，即 $i = 1, 2, 3, \cdots, n$；

n——总层数；

E_V——纵向波及系数；

h_t——总厚度；

h_i——小层厚度。

若各层孔隙度值不同，则公式为：

$$E_V = \frac{\left(\frac{K}{\phi}\right)_i \sum_{j=1}^{i} (\phi h)_j + \sum_{j=i+1}^{n} (Kh)_j}{\left(\frac{K}{\phi}\right)_i \sum_{j=1}^{i} (\phi h)_j} \tag{3-30}$$

Stiles 还提出了以下公式，用于计算在各小层见水时的地面水油比：

$$WOR_s = A \frac{\sum_{j=1}^{i} (Kh)_j}{\sum_{j=i+1}^{n} (Kh)_j} \tag{3-31}$$

其中

$$A = \frac{K_{rw}\mu_o B_o}{K_{ro}\mu_w B_w} \tag{3-32}$$

可以同时使用纵向波及系数和地面水油比来描述从第 1 层到第 n 层按顺序见水动态。通常将计算结果做成水油比（WOR）随 E_V 变化的关系曲线，这样更加直观。

2. Dykstra – Parsons 法

Dykstra 和 Parsons（1950）将纵向波及系数与以下参数建立了联系：渗透率变异系数 V、流度比 M 和地下水油比 WOR_r，给出了水油比为 0.1、0.2、0.5、1、2、5、10、25、50 和 100 相应的关系图版，图 3-14 为 $WOR=50$ 的纵向波及系数图版。De Souza 和 Brigham（1981）利用回归分析模型，将 $0 \leq M \leq 10$、$0.3 \leq V \leq 0.8$ 范围的纵向波及系数曲线回归成一条曲线，如图 3-15 所示。用 WOR、V 和 M 综合定义图 3-15 里面的相关参数 Y：

$$Y = \frac{(WOR + 0.4)(18.948 - 2.499V)}{(M - 0.8094V + 1.137)10^x} \tag{3-33}$$

其中

$$x = 1.6453V^2 + 0.935V - 0.6891 \tag{3-34}$$

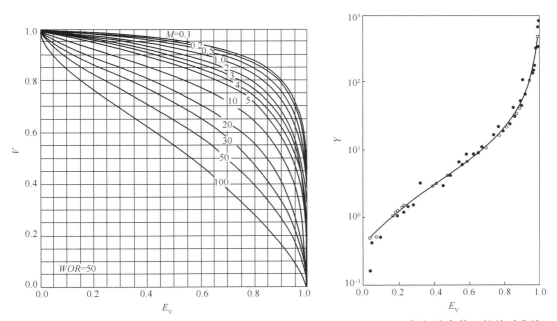

图 3-14　$WOR=50$ 的纵向波及系数图版　　图 3-15　E_V 与相关参数 Y 的关系曲线

计算纵向波及系数与水油比关系曲线的具体步骤如下：

(1) 计算流度比 M 和渗透率变异系数 V。

(2) 给定若干个 WOR 值，例如 1、2、5、10 等，计算给定 WOR 的相关参数 Y。

(3)由图3-15确定对应纵向波及系数E_V值。

(4)绘制E_V与WOR的关系曲线

为进一步简化E_V的计算过程,Fassihi(1986)对图3-15的曲线进行了拟合,给出了如下的非线性方程。利用该方程,通过迭代求解,可以得到纵向波及系数E_V:

$$a_1 E_V^{a_2}(1-E_V)^{a_3} - Y = 0 \qquad (3-35)$$

式中,$a_1 = 3.334088568$,$a_2 = 0.7737348199$,$a_3 = 1.225859406$。

Newton-Raphson方法是求解式(3-35)比较合适的方法。为了避免迭代过程,可以使用以下公式计算纵向波及系数:

$$E_V = a_1 + a_2 \ln Y + a_3 (\ln Y)^2 + a_4 (\ln Y)^3 + a_5/\ln Y + a_6 Y$$

式中,$a_1 = 0.19862608$,$a_2 = 0.18147754$,$a_3 = 0.01609715$,$a_4 = -4.6226385 \times 10^{-3}$,$a_5 = -4.2968246 \times 10^{-4}$,$a_6 = 2.7688363 \times 10^{-4}$。

第三节 多层油藏开发动态预测方法

在预测开发动态时,为了考虑油藏纵向非均质性,假定多层油藏层间没有纵向连通,即层与层之间不存在窜流。小层属性参数包括厚度h、渗透率K和孔隙率ϕ。整个油藏的非均质性通常用渗透率变异系数V来表征。下面讨论三种用于预测多层油藏开发动态的方法。

一、简化的 Dykstra-Parsons 方法

Dykstra 和 Parsons(1950)提出了一个方法,以流度比、渗透率变异系数和生产水油比作为相关参数预测水驱采出程度。基于 Dykstra-Parsons 的方法,预测水油比为1、5、25 和100 对应的总采出程度R,基于这些数据,Johnson(1956)建立了简化图版。图3-16 为四个WOR_s对应的图版,图中每条曲线对应一个$R(1-S_w)$值。

简化的 Dykstra-Parsons 方法的实际应用步骤如下:

(1)计算渗透率变异系数V和流度比M。

(2)利用渗透率和流度比,由WOR_s为1、5、25、100 对应的四个图版计算总采出程度R。例如,在V和M分别为0.5 和2 情况下,求WOR_s为5 时的采出程度:

①在相应图版上,找$V=0.5$和$M=2$对应的曲线;

②二者的交点对应的$R(1-0.72S_{wi})=0.25$;

③若初始含水饱和度S_{wi}为0.21,则可求出采出程度为$R=0.29$。

(3)计算四个水油比(即1、5、25 和100)对应的累积产油量N_p。

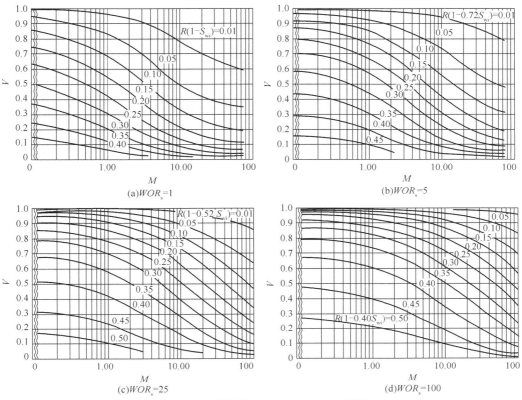

图 3-16 简化的 Dykstra 和 Parsons 图版

（4）在半对数纸或直角坐标纸上绘制水油比与采出程度关系曲线，将这条线外推至非常低 WOR_s 值，即可得到见水时采出程度，如图 3-17 所示。

（5）在恒速注入情况下，填充注入体积 W_{if} 加上见水时累积产油量，除以注入速度即可估算见水时间。

（6）对于某一 WOR_s 值，WOR_s 与 N_p 关系曲线所包围面积，等于其对应的累积产水量，如图 3-18 所示。

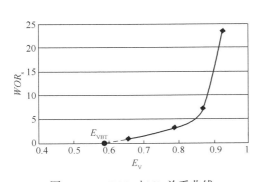

图 3-17 WOR_s 与 E_V 关系曲线

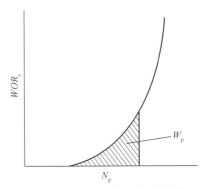

图 3-18 WOR_s 与 N_p 关系曲线

(7) 对于某一 WOR_s 值,其对应的累积注水量等于其对应的累积产油量和累积产水量再加上填充体积之和:

$$W_{inj} = N_p B_o + W_p B_w + W_{if}$$

二、改进的 Dykstra – Parsons 方法

Felsenthal、Cobb 和 Heuer(1962)对 Dykstra – Parsons 的方法进行了改进,考虑了水驱起始时初始含气饱和度这一因素。假设注水量 i_w 恒定,该方法的步骤如下:

步骤 1:进行基础性计算,确定如下信息:(1)孔隙体积 PV 和原始地质储量 N_S;(2)含水率 f_w 与含水饱和度 S_w 的关系数据或者函数;(3)$(\mathrm{d}f_w/\mathrm{d}S_w)$ 与 S_w 的关系函数或者关系数据;(4)见水时平均含水饱和度 $\bar{S}_{wBT}$;(5)流度比 M;(6)见水时驱油效率 E_{DBT};(7)见水时平面波及系数 E_{ABT};(8)渗透率变异系数 V;(9)填充注入体积 W_{if}。

步骤 2:计算水油比 WOR 为 1、2、5、10、15、20、25、50 和 100 对应的纵向波及系数。

步骤 3:在笛卡儿坐标系绘制 WOR 与 E_V 关系曲线,将 WOR 与 E_V 关系曲线外推至 $WOR = 0$,确定见水时纵向波及系数 E_{VBT}。

步骤 4:计算见水时累积注水量:

$$W_{iBT} = PV(\bar{S}_{WBT} - S_{wi})E_{ABT}E_{VBT}$$

步骤 5:计算见水时累积产油量:

$$(N_p)_{BT} = \frac{W_{iBT} - W_{if}E_{VBT}}{B_o}$$

步骤 6:计算见水时间 t_{BT}:

$$t_{BT} = \frac{W_{iBT}}{i_W}$$

步骤 7:给定若干个地下水油比 WOR_r 值。

步骤 8:确定各 WOR_r 值对应的 E_V 值(参见步骤 3)。

步骤 9:将给定的 WOR_r 值分别转换为生产井井底含水率 f_{w2} 和地面水油比 WOR_s:

$$f_{w2} = \frac{WOR_r}{WOR_r + 1}$$

$$WOR_s = WOR_r \left(\frac{B_o}{B_w}\right)$$

步骤 10:根据含水率和含水饱和度关系曲线确定每个 f_{w2} 值对应的含水饱和度 S_{w2}。

步骤 11:确定每个 f_{w2} 值对应的平面波及系数 E_A。

步骤 12:确定每个 f_{w2} 值对应的纵向波及系数 E_V。

步骤 13:确定每个 f_{w2} 值的平均水饱和度 $\bar{S}_{w2}$。

步骤 14：计算步骤 13 中每个 $\bar{S}_{w2}$ 值对应的驱油效率 E_D。

步骤 15：计算每个 WOR_s 的累积产油量：

$$N_p = N_S E_D E_A E_V - \frac{PVS_{gi}(1 - E_A E_V)}{B_o}$$

步骤 16：在笛卡儿坐标系上画出累积产油量 N_p 与 WOR_s 的关系曲线，求出各个 WOR_s 值对应的曲线包围面积，该面积表示给定的 WOR_s 值对应的累积产水 $(W_p)_{WOR}$。

步骤 17：计算各 WOR_s 值对应的累积注水量 W_{inj}：

$$W_{inj} = (N_p)_{WOR} B_o + PVS_{gi}(E_V)_{WOR}$$

式中　$(N_p)_{WOR}$——水油比达到 WOR 时的累积产油量；

　　　$(E_V)_{WOR}$——水油比达到 WOR 时的纵向波及系数。

步骤 18：计算 W_{inj} 对应的时间：

$$t = \frac{W_{inj}}{i_w}$$

步骤 19：分别计算产油量和产水量：

$$Q_o = \frac{i_w}{B_o + B_w WOR_s}$$

$$Q_w = Q_o WOR_s$$

三、Craig – Geffen – Morse 方法

在计算开采动态时，如果融入纵向波及系数的计算，就会使得难度明显增大。所以，Craig 等人(1955)提出了一种新的方法。对于多层油藏，从其中选择一个层，对这一个层开展相关计算。选定的这个层，称为基准层，认为其纵向波及系数为 100%，其他层的动态可以通过"平移时间刻度"得到。流程如下：

步骤 1：将油藏划分为合适数目的小层。

步骤 2：选定基准层(例如，第 n 层)，计算该小层的动态。

步骤 3：绘制基准层(第 n 层)累积量(N_p、W_p、W_{inj})和产量、注水量(Q_o、Q_w、i_w)随时间 t 的变化曲线。

步骤 4：对每一层(包括基准层)计算 K/ϕ、ϕh 和 Kh。

步骤 5：为了得到第 i 层的动态，给定一系列时间 t，用下式计算 t^*，从步骤 3 得到的曲线上确定相应的 N_p^*、W_p^*、W_{inj}^*、Q_o^*、Q_w^* 和 i_w^*：

$$t_i^* = t \frac{\left(\dfrac{K}{\phi}\right)_i}{\left(\dfrac{K}{\phi}\right)_n}$$

然后计算第 i 层在任意时刻 t 的动态：

$$N_\mathrm{p} = N_\mathrm{p}^* \frac{(\phi h)_i}{(\phi h)_n}$$

$$W_\mathrm{p} = W_\mathrm{p}^* \frac{(\phi h)_i}{(\phi h)_n}$$

$$W_\mathrm{inj} = W_\mathrm{inj}^* \frac{(\phi h)_i}{(\phi h)_n}$$

$$Q_\mathrm{o} = Q_\mathrm{o}^* \frac{(K/\phi)_i}{(K/\phi)_n}$$

$$Q_\mathrm{w} = Q_\mathrm{w}^* \frac{(K/\phi)_i}{(K/\phi)_n}$$

$$i_\mathrm{w} = i_\mathrm{w}^* \frac{(K/\phi)_i}{(K/\phi)_n}$$

步骤6：整体水驱动态由各层加和得到。

第四章 油田动态分析的经验方法

经验方法的一般工作程序就是系统地观察油田的生产动态,准确齐全地收集能说明生产规律的资料,深入地分析这些资料,以发现其中带规律性的东西,进而对这些带规律性的资料和数据进行数学处理,并给出表达这些规律的经验公式(包括其中经验参数的确定),最终运用已经总结出来的经验规律,对今后的生产动态进行预测。

第一节 油田产量递减规律

无论何种储集类型,也无论何种驱动类型的油气田,随着开发的深入发展,都会进入产量递减阶段。根据该阶段的产量和累积产量的数据,利用产量递减分析法,既可预测油田未来时间的产量和累积产量的变化,又可对油田的可采储量和剩余可采储量作出有效的预测。本节将介绍油田产量递减规律分析方法,以及递减规律的判别与应用。

一、产量变化模式

油田产量变化一般分三个阶段:上产期、稳产期和递减期,如图 4-1 所示。

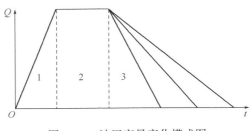

图 4-1 油田产量变化模式图
1—上产期;2—稳产期;3—递减期

1. 上产期

油田投入开发的初期,新井不断投产,生产设施也不断完善,所以产量也逐年上升,这个阶段称作油气田的产量上升期,简称上产期。上产期通常比较短,2~5年不等,具体时间与油田规模、地质条件、政治、经济及技术条件有关。由于该阶段只能采出可采地质储量的5%~10%,且时间短,主要受人为因素的干扰,因此上产期的产量变化规律一般很少有人研究。

2. 稳产期

上产期结束后,油田开发井网、管网系统建设已基本完成,油田开发的整体生产设施也已基本建立起来,油气水分离和集输系统已开始正产运转,注入系统也开始正常工作,之后的一段时间,油气生产将进入满负荷运转期,油气产量也达到最大设计能力,这一阶段称作油气田开发的稳产期。稳产期是油气田开发的黄金时期,该阶段的长短主要受油气藏地质条件和开发系统设置的影响。一般情况下,储量规模越大,稳产期越长。按照行业规定,中小型油田的稳产期应在2~5年;大中型油田的稳产期应在5~10年;特大型油田的稳产期一般都在10年以上(例如大庆油田稳产5000×10^4t持续了27年)。一个油气田开发的稳产期到底多长合适,最终要通过经济效益最大化原则对其进行设计。稳产期的开发效益最好,一般能采出可采地质储量的50%左右。其长短与产量有关,稳产期的时间T满足以下公式:

$$T = \frac{NR_o}{Q}$$

式中　N——地质储量;

　　　R_o——稳产期采出程度;

　　　Q——稳产期的产量。

由上式可以看出,油气田稳产期的产量越高,稳产期越短。

3. 递减期

稳产期结束之后,油田开发进入产量递减期。递减期是任何一个油气田都无法回避的开采阶段。递减期的出现,是地层有效驱动能量正趋于衰竭的标志。靠天然能量驱动的油田,经过稳产期的大幅度开采,地层能量已消耗殆尽,只有降低产量以求得低驱动能量水平上的新平衡;注水开发或天然水驱的油田,经过稳产期的开采,大部分油井都已见水,产水后,水的驱动效率降低,油产量随之递减。与上产期和稳产期不同的是,递减期产量变化是一个自然的过程。但由于油气田地质条件的差异和开发系统设置上不同,产量递减路径和递减模式各不相同(图4-1)。这就需要一定的研究工作,找出特定油田的特定递减规律和影响油气产量的主要原因及矛盾,以便对油气生产作出预测和规划,提高油气田开发的经济效益。

递减期的长短主要受油气田地质条件和当时的经济技术条件影响,大多数油田都有一个长长的产量递减期,一般都在10~30年以上。递减期可以采出可采地质储量的40%~50%。

由于递减期的油气产量不断减小,油田经济效益不断下滑。为提高经济效益,一般情况下会采取以下措施减缓产量递减:细分开发层系、井网调整、注采系统调整、EOR方法、增产增注措施等。待所有实施完毕后,油气生产仍不能带来经济效益,油气开采过程将被终止,油田最终被废弃。

二、油田产量递减规律研究方法

研究产量递减规律的方法一般是:首先绘制产量与时间的关系曲线,或产量与累积产量的关系曲线,然后选择一种坐标将产量递减部分的曲线变成直线(或接近直线),写出直线的方程,即找出油田产量与时间的经验关系式,进而预测油田未来的动态指标。油田产量递减规律矿场上常见的有指数递减规律和双曲线型的衰减规律。

1. 产量递减率

在油田开发过程中,随着地下能量的变化和可采储量的减小,产油量总是要下降的,通常用递减率表示产量的递减速度。所谓递减率,是指单位时间内单位产量的变化,通常用小数或百分数来表示。根据矿场实际资料统计分析,可以把递减率表达为下面的形式:

$$D = -\frac{\mathrm{d}q}{q\mathrm{d}t} = kq^n \tag{4-1}$$

式中　D——产量递减率,单位为时间单位的倒数;

　　　q——产量;

　　　t——时间;

　　　n——递减指数,$0 \leq n \leq 1$;

　　　k——比例常数。

式(4-1)中的负号表示随着开发时间的增长,产量是下降的。

[**例题4-1**]　请根据递减率公式,利用表4-1中所给数据,计算该井组2020年的自然递减率和综合递减率。

表4-1　井组开发数据表

时间	年产液量(t)	年产油量(t)	措施增油量(t)
2019年	83976	7222	126
2020年	100450	6630	94

解:在计算该井组2020年的自然递减率时需要将年产油量的增量扣除94t的措施增油量,所以有:

自然递减率 = [7222 - (6630 - 94)] ÷ 7222 = 9.5%(a^{-1})

综合递减率 = (7222 - 6630) ÷ 7222 = 8.2%(a^{-1})

总结：实际上，矿场上经常会用到递减率来刻画产量递减的快慢。一般情况下，当 $D < 0.1a^{-1}$ 时，产量递减较慢；当 $D = 0.1 - 0.3a^{-1}$ 时，产量递减中等；当 $D > 0.3a^{-1}$ 时，产量递减较快。

产量递减的快慢，受多种因素影响，每个油藏的影响因素又各有不同。一般情况下，单井控制储量大小、天然能量补给速度和含水上升速度是影响产量递减的主要客观因素，采油速度和人工能量补给速度是影响产量递减的主要人为因素。

2. 产量递减规律的相关公式

产量递减规律类型主要有双曲线型递减，指数型递减和调和型递减三种，相关公式见表 4-2。

表 4-2 产量递减规律的相关公式

递减类型	基本特征	基本关系式			最大累积产量
		$q_t - t$	$N_p - t$	$N_p - q_t$	
指数型递减	$n = 0$ $D = D_i$	$q_t = q_i \exp(-D_i t)$	$N_p = \dfrac{q_i}{D_i}[1 - \exp(-D_i t)]$	$N_p = \dfrac{1}{D_i}(q_i - q_t)$	$N_p = \dfrac{q_i}{D_i}$
双曲线型递减	$0 < n < 1$ $D < D_i$	$q_t = q_i(1 + nD_i t)^{-\frac{1}{n}}$	$N_p = \dfrac{q_i}{(n-1)D_i}$ $\left[(1 + nD_i t)^{\frac{n-1}{n}} - 1\right]$	$N_p = \dfrac{q_i^n}{(1-n)D_i}$ $\left[q_i^{1-n} - q_t^{1-n}\right]$	$N_p = \dfrac{q_i}{(1-n)D_i}$
调和型递减	$n = 1$ $D < D_i$	$q_t = q_i(1 + D_i t)^{-1}$	$N_p = \dfrac{q_i}{D_i}\ln(1 + D_i t)$	$N_p = \dfrac{q_i}{D_i}\ln\dfrac{q_i}{q_t}$	$N_p = \dfrac{q_i}{D_i}\ln q_i$

注：D_i—初始递减率；q_i—初始产量；q_t—第 t 年的产量；n—递减指数。

3. 三种递减规律的比较

（1）当 $n = 0$，$D_t = D_i =$ 常数，即指数型递减时的递减率为常数，也称为常百分数递减。

（2）当 $n = 1$，$D_t = D_i \dfrac{q_t}{q_i}$，由于 $\dfrac{q_t}{q_i} < 1$，所以调和型递减时的递减率随着产量的下降而减小，即随着开发时间的推移，递减速度逐渐减缓。

（3）当 $0 < n < 1$ 时，$D_t = D_i \left(\dfrac{q_t}{q_i}\right)^n$，因为 $\left(\dfrac{q_t}{q_i}\right)^n < 1$，所以双曲线型递减时的递减率也随着产量的下降而减小。

根据上面的推证得知，在初始递减率相同的条件下，指数型递减产量下降最快。

由于 $\dfrac{q_t}{q_i} < 1$ 且 $0 < n < 1$，有 $\left(\dfrac{q_t}{q_i}\right)^n > \dfrac{q_t}{q_i}$，可以推知：双曲线型递减时的递减率要比调和型的递减率大，即在初始递减率相同的条件下，调和型递减的产量下降幅度要比双曲线型的产量下降幅度小。

综上所述：产量递减速度主要决定于递减指数 n 和初始递减率 D_i。在初始递减率和递减期初始产量相同时，指数型递减产量下降最快，双曲线型递减居中，调和型递减最慢。在递减类型一定时,初始递减率越大,产量下降越快。三种递减类型的典型曲线如图 4-2 所示。

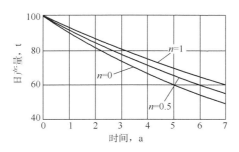

图 4-2 典型递减规律曲线

4. 产量递减类型的判别

此处主要介绍四种判别方法:图解法、试凑法(试差法)、典型曲线拟合法、诊断曲线法。

1) 图解法

(1) 做 $\lg q_t$—t 曲线,若为直线,则服从指数型递减,如图 4-3 所示。

(2) 做 $\lg q_t$—N_p 曲线,若为直线,则服从调和型递减,如图 4-4 所示;否则,服从双曲线型递减。

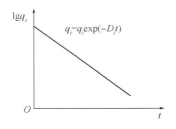
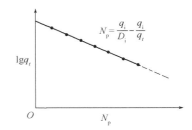

图 4-3 指数型产量递减规律　　　　图 4-4 调和型产量递减规律

2) 试凑法(也叫试差法)

当产量递减规律不满足指数型和调和型递减规律时,则只能是双曲线型产量递减规律。但该方法直接线性回归无法确定其参数,只能用试凑法(或试差法)求解和判断。

由表 4-2 中的双曲线型产量与时间公式变形得:

$$\left(\frac{q_i}{q_t}\right)^n = a + bt \tag{4-2}$$

其中　　　　　　　　　　$a = 1, \quad b = nD_i$

先人为给定 n 值,如果数据回归满足很好的线性关系,则 n 值为准确的;若曲线上翘,则 n 值偏大;反之,曲线向下偏移(即向横轴偏),则 n 值偏小,如图 4-5 所示。调整 n 值,直到数据满足线性关系,从而确定 n 值和 a、b 值,进而求解出 D_i 值。

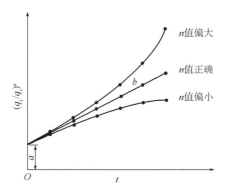

图 4-5 双曲线型递减规律的试算法图示

3) 典型曲线拟合法

根据表 4-2 中产量与时间关系式,变形得:

$$\frac{q_\mathrm{i}}{q_t} = \exp(D_\mathrm{i} t) \quad (n=0,\text{指数型递减})$$

$$\frac{q_\mathrm{i}}{q_t} = (1+nD_\mathrm{i} t)^{\frac{1}{n}} \quad (0<n<1,\text{双曲线型递减})$$

$$\frac{q_\mathrm{i}}{q_t} = 1 + D_\mathrm{i} t \quad (n=1,\text{调和型递减})$$

该方法的具体步骤如下:

(1) 做理论图版(给定不同 n 和 $D_\mathrm{i} t$ 值,做 q_i/q_t 与 $D_\mathrm{i} t$ 曲线),如图 4-6 所示;

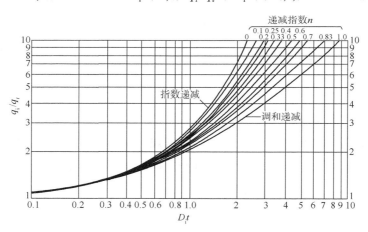

图 4-6 典型曲线拟合法理论图版

(2) 做实际曲线(递减阶段的 q_i/q_t 与 t 关系曲线,透明纸);

(3) 拟合,得到 n 和 D_i 值。

其中,n 可以直接读出,D_i 值通过实际曲线与理论图版重合的某一数据点计算得出,计算公式如下:

$$D_i = \frac{D_i t(理论)}{t(实际)} \tag{4-3}$$

4）诊断曲线法

产量的递减率一般不是常数，而是随产量变化的一个变量。根据大量的统计规律发现，递减率满足：

$$D = Kq^n \tag{4-4}$$

对其两边取对数，得：

$$\lg D = \lg K + n \lg q \tag{4-5}$$

根据递减率的定义式 $D = -\frac{\mathrm{d}q}{q\mathrm{d}t}$，求出每个产量对应的递减率后，再按照式（4-5）在双对数坐标系中绘制递减率与产量之间的关系曲线，如图 4-7 所示。由图中曲线可以看出，在产量变化的某个阶段，递减率与产量的双对数曲线为一直线，该直线的斜率即为递减指数 n。同一直线的产量点满足同一递减类型。李传亮把该曲线称作产量递减规律的诊断曲线，把式（4-5）称为产量递减规律的诊断方程。

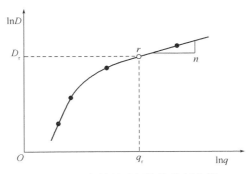

图 4-7　产量递减规律的诊断曲线

第二节　含水上升规律

对于水驱油田来说，无论是依靠人工注水还是依靠天然水驱采油，在无水采油期结束以后，将长期地进行含水生产，其含水率还将逐步上升，这是影响油田稳产的重要因素。水驱油田的含水率是油田开发中受多方面因素影响的一个综合指标，它既反映油层及原油物性对油层中油水运动规律的制约，也反映开采过程中多种技术措施的效果。因此，对这类油田，利用油田开采中的实际生产资料，分析认识含水上升规律，研究影响含水上升规律的地质工程因素，制定不同生产阶段的切实可行的控水稳油、控水增油的措施，是开发水驱油田的一件经常性的、极为重要的工作，本节将主要介绍水驱油田含水上升规律及其应用。

一、油田含水率变化规律

通常,表示水驱油田开发动态的一个基本曲线是含水率与采出程度关系曲线,如图4-8所示。这一曲线的形态及位置,综合反映了储层地质特征、油水分布及性质、开发方式及工艺措施的水平。从研究工作的角度看,这条曲线是一条形状较为特殊的曲线,难以用简单的公式来表达,所以研究含水规律时,需要对这些经验数据进行一定的数学处理和变换。

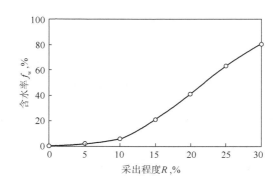

图4-8 某油田含水率与采出程度关系曲线

生产实践表明,一个天然水驱或人工水驱的油藏,当它已全面开发并进入稳定生产以后,其含水率达到一定程度并逐步上升时,在半对数坐标纸上,以累积产水量的对数为纵坐标,以累积产油量(或者采出程度)为横坐标,二者的关系曲线是一条直线,该曲线称为水驱规律曲线。

图4-9是我国某油田注水开发的水驱曲线。这条直线一般从中含水期开始(含水率在20%左右)出现,而到高含水期仍保持不变。在油田的注采井网、注采层系、注采强度等开采方式保持不变时,直线性质始终保持不变;当注采方式变化后,则出现拐点,但直线关系仍然成立。如图4-9中在含水率达到47.7%左右时,直线出现了拐点,这是因为此时采取了一定调整措施。

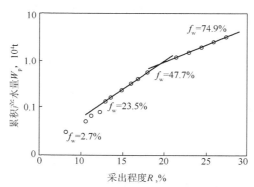

图4-9 某油田调整措施前后水驱规律曲线的变化

发现这一规律十分重要,因为有了这样的规律就可以将油田含水规律正确地表达出来。油水产量之间的这种半对数关系在国内外许多油田上都可以看到,具有相当广泛的普遍性。在我国的注水开发油田当中,绝大部分也都符合这种规律。这样,人们就可以运用这一定量规律来描述和预测各油田在生产过程中的含水变化、产油水情况、最终采收率及可采储量等。

目前一致的观点是:不同油水黏度的油田水驱规律曲线有明显差异,对低黏度油田,油水黏度比低,开发初期含水率上升缓慢,在含水率与采出程度的关系曲线上呈凹型曲线,主要储量在中低含水期采出;而中高黏度油田与此相反,在含水率与采出程度的关系曲线上呈凸形曲线,主要储量在高含水期采出,这是由水驱油的非活塞性所决定的。储层的润湿性和非均质性更加剧了这种差异。

目前一般含水率划分标准如下:
(1)无水采油期:含水率小于2%;
(2)低含水采油期:含水率为2%－20%;
(3)中含水采油期:含水率为20%－60%;
(4)高含水采油期:含水率为60%－90%;
(5)特高含水采油期:含水率大于90%;
(6)极限含水采油期:含水率等于98%。

二、常见水驱规律曲线

对于水驱规律曲线的研究和应用大致经历了三个阶段,各阶段都有其典型的研究方法。

1. 含水与时间或含水与采出程度的关系曲线

1959 年,苏联学者马克西莫夫在确定水驱油藏末期的可采储量时,用格罗兹内油区一些老油田资料进行研究,认为在水驱油田末期对一个层系而言,累积产油量 N_p 和累积产水量 W_p 之间存在着一种统计关系,其表达式为:

$$W_p = a_1 \exp(b_1 N_p) \tag{4-6}$$

式中　a_1, b_1——与油田的地质、开发等因素有关的待定系数。

1971 年,美国学者蒂麦尔曼统计了美国一些水驱油田的实际资料,指出油水比 F_{ow} 与累积产油量 N_p 在半对数坐标系中是一直线关系,其表达式为:

$$\lg F_{ow} = b_2 N_p + a_2 \tag{4-7}$$

其中　　　　　　　　　　　$F_{ow} = q_o / q_w$

式中　F_{ow}——油水比;
　　　a_2, b_2——常数。

他们的研究指出了油、水量的变化有一定的定量关系,但对其应用则研究得不够充分。因此在这一阶段,矿场上仍用老方法,做含水率 f_w 与时间 t,或含水 f_w 与采出程度 R 的关系曲线。这种做法简便直观,直到今天矿场仍在采用,但由于影响因素较多,实际资料往往波动较大,规律性不强。

2. 水驱规律曲线

进入 20 世纪 70 年代以来,随着油田普遍进行注水开发,很多油藏动态的研究人员发现,一个天然水驱或人工水驱的油藏,当它已经全面开发并进入稳定生产阶段后,含水率达到一定高度并逐渐上升时,在半对数坐标纸上,以对数坐标表示油藏的累积产水量 W_p,以普通坐标表示油藏的累积产油量 N_p,做出两者的关系曲线,常出现一条近似的直线段,如图 4-10 所示。这类曲线,称为水驱规律曲线(后来也称甲型水驱规律曲线),其基本表达式有如下两种:

$$\lg W_p = BN_p + A \tag{4-8}$$

$$R = B\lg\left(\frac{f_w}{1-f_w}\right) + A \tag{4-9}$$

式中　A,B——水驱常数;

　　　R——采出程度;

　　　f_w——含水率。

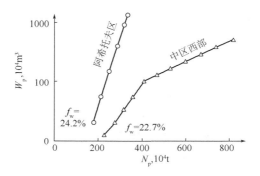

图 4-10　累积产水量与累积产油量关系曲线

在实际工作中,人们称这种方法为驱替特征法。进一步细分,式(4-8)为累积产水量—累积产油量关系曲线法;式(4-9)为采收率—水油比关系曲线法。两式本质相同,并可相互推证。

水驱规律曲线有下述两大特点:

(1)一般油田在含水 20% 前后开始出现直线段。

(2)在对油层采取重大措施(如压裂)或开发条件变动(如层系调整)时,开发效果突变,直线段发生转折。

目前,人们在矿场上利用水驱规律曲线判断或对比油田(或油井)开采效果的好坏,判断油井出水层位及来水方向或见效方向,以及预测油藏动态指标。

3. 驱替系列公式

基于我国油田普遍采用注水开发方式,考察不同储油物性油藏的驱替特征发现,驱替特征法对岩石和流体物性中等的油田是适用的,并且含水率随采出程度变化规律呈 S 型,而对岩石和流体物性很好或较差油田不太适用或很不适用。实践证明,各种性质的油田具有不同类型的驱替特征形态。我国石油工程师万吉业论述了含水率—采收率关系曲线与油田储油层岩石孔隙结构、流体性质及其润湿性的关系,并根据其特点可分为五种驱替特征类型(图 4 – 11),简称驱替系列,用公式表达为:

(1) 凸型曲线:

$$R = A + B\ln(1 - f_w) \tag{4 – 10}$$

(2) 凸型和 S 型间过渡曲线:

$$\ln(1 - R) = A + B\lg(1 - f_w) \tag{4 – 11}$$

(3) S 型曲线:

$$R = A + B\ln\frac{f_w}{1 - f_w} \tag{4 – 12}$$

(4) S 型和凹型间过渡曲线:

$$\ln R = A + B f_w \tag{4 – 13}$$

(5) 凹型曲线:

$$\ln R = A + B\ln f_w \tag{4 – 14}$$

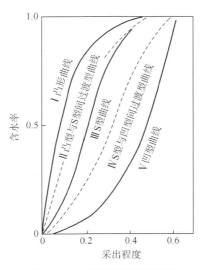

图 4 – 11 五种驱替特征曲线

4. 其他统计关系

研究油田水驱曲线的数学模型还有很多,实际应用时都可以进行尝试,甚至可以设计出一些新的数学关系式,下面是几种常见的统计关系曲线方程。

(1)方程1(乙型水驱规律曲线):

$$\ln L_p = a + b N_p \qquad (4-15)$$

其中
$$L_p = W_p + N_p$$

式中 L_p——累积产液量。

(2)方程2(丙型水驱规律曲线):

$$\frac{L_p}{N_p} = a + b L_p \qquad (4-16)$$

(3)方程3(丁型水驱规律曲线):

$$\frac{L_p}{N_p} = a + b W_p \qquad (4-17)$$

(4)方程4:

$$\ln R_{wo} = a + b N_p \qquad (4-18)$$

式中 R_{wo}——生产水油比。

(5)方程5:

$$\ln R_{Lo} = a + b N_p \qquad (4-19)$$

式中 R_{Lo}——生产液油比,即产液量与产油量的比值。

(6)方程6:

$$\ln f_w = a + b N_p \qquad (4-20)$$

(7)方程7:

$$\ln f_w = a + b \ln N_p \qquad (4-21)$$

三、甲型水驱规律曲线及其应用

1. 理论推导

甲型水驱规律曲线应用时,式(4-8)可变形为:

$$N_p = a(\lg W_p - \lg b) \qquad (4-22)$$

如前所述,水驱油田开发到一定阶段以后,在半对数坐标系下,以累积产油量为普通坐标,以累积产水量为对数坐标,绘制出水驱规律曲线为一直线,如图4-12所示,其斜率可由下式求得:

$$\frac{1}{a} = \frac{\lg W_{p2} - \lg W_{p1}}{N_{p2} - N_{p1}} \qquad (4-23)$$

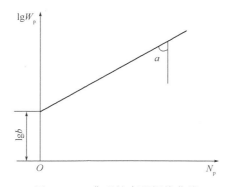

图 4-12 典型的水驱规律曲线

将直线段延长与纵轴相交,得截距 b。求出常数 a,b 后可进行开发动态预测。

1) 预测产水量

将式(4-8)变形为

$$W_p = b \cdot 10^{(N_p/a)} \tag{4-24}$$

或者

$$W_p = 10^{(\lg b + N_p/a)} \tag{4-25}$$

在给定产量 N_p 的条件下,根据式(4-24)或式(4-25)即可求得产水量 W_p。

2) 预测含水率

对式(4-22)取微分有:

$$\frac{dN_p}{dt} = a \lg e \frac{1}{W_p} \frac{dW_p}{dt}$$

令 $\frac{dN_p}{dt} = q_o, \frac{dW_p}{dt} = q_w$,代入上式整理得水油比 F_{wo} 为:

$$F_{wo} = \frac{q_w}{q_o} = \frac{2.3 W_p}{a} \tag{4-26}$$

含水率可根据下式求得:

$$f_w = \frac{q_w}{q_w + q_o} = \frac{1}{1 + q_o/q_w}$$

将式(4-26)代入上式得:

$$f_w = \frac{2.3 W_p}{a + 2.3 W_p} \tag{4-27}$$

根据式(4-27)即可预测含水率。

3) 预测最终采收率

目前在水驱油田上普遍采用含水极限或极限水油比这一概念,超过了这一极限,油田就失去了实际开采价值。达到这一极限所获得的采出程度就是油田的最终采收率。一般通用的含水极限为 98% 或极限水油比为 49。经验方法所预测的采收率值一般比其他方法更符合生产实际。

将式(4-26)变形有：

$$W_p = \frac{aF_{wo}}{2.3}$$

代入式(4-22)得：

$$N_p = a\left(\lg\frac{aF_{wo}}{2.3} - \lg b\right) \tag{4-28}$$

将极限水油比 $F_{wo} = 49$ 代入式(4-28)得：

$$N_{pmax} = a(\lg 21.3a - \lg b) \tag{4-29}$$

生产中分析油田开发效果时，往往要求知道含水与采出程度的关系，以便于对比，为此取下面三个无量纲量，即：

$$R = N_p/N$$
$$a_D = a/N$$
$$b_D = b/N$$

式中 N 为地质储量，则有：

$$R_{max} = \frac{N_{pmax}}{N} = \frac{a}{N}(\lg 21.3a - \lg b) \tag{4-30}$$

或者

$$R_{max} = a_D(\lg 21.3a_D - \lg b_D) \tag{4-31}$$

根据上式所求得的最大采出程度 R_{max} 即为油田的最终采收率。

2. 实际应用

[例 4-2] 已知某油田 L 油层实施注水开发，其生产数据见表 4-3。油田地质储量为 $2409 \times 10^4 t$。试根据水驱规律曲线计算与累积采油量对应的含水率，并求含水率达 98% 时的最终采油量和采收率。

表 4-3 某油田 L 层生产数据及计算结果表

序号	时间年份	生产数据			计算数据		
		累积产油量 (10^4t)	累积产水量 ($10^4 m^3$)	含水率 (%)	累积产水量 ($10^4 m^3$)	含水率 (%)	采出程度 (%)
1	1959	388.7	32.9	20.8	42.7	29	16.09
2	1960	434.4	58.2	35.1	66.0	38.8	18.03
3	1961	465.7	82.0	43.3	89.0	46	19.33
4	1962	490.2	105.7	49.0	112	49	20.43

续表

序号	时间年份	生产数据			计算数据		
		累积产油量 (10^4t)	累积产水量 (10^4m³)	含水率 (%)	累积产水量 (10^4m³)	含水率 (%)	采出程度 (%)
5	1963	508.9	125.6	51.3	135	56	21.12
6	1964	526.9	151.0	58.6	159	60	21.87
7	1965	544.1	180.9	63.4	186	65	22.59
8	1966	562.4	214.5	64.9	224	68	23.34
9	1967	581.1	249.2	64.9	269	72	24.12
10	1968	599.9	281.4	63.0	282	62.5	24.90
11	1969	621.2	316.6	62.3	316	65	25.79
12	1970	644.1	360.7	65.8	363	68	26.74
13	1971	666.3	407.7	68.0	417	71	27.66
14	1972	687.6	460.4	71.3	468	73	28.54
15	1973	709.4	527.6	75.5	525	76.6	29.44
16	1974	730.1	600.3	77.8	600	78	30.30
17	1975	751.9	696.9	81.6	693	80	31.21
18	1976	762.6	729.1	83.5	725	81	31.65

解:(1)根据给出的累积产油量 N_p 和累积产水量 W_p,在半对数坐标纸上绘制出水驱规律曲线,如图 4-13 所示。

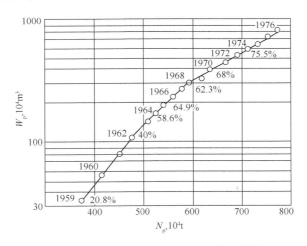

图 4-13 累积产水和累积产油关系曲线

(2)确定描述直线段的常数 a 和 b,给出直线方程。从图中看到,直线在 1968 年出现拐点,因此需要分段进行计算。

1968 年以前:

$$\frac{1}{a_1} = \frac{\lg 1000 - \lg 100}{720 - 480} = \frac{1}{240}$$

$$a_1 = 240$$

查图得：$b_1 = 1.023$，$\lg b_1 = 0.01$，因此第一直线段方程为：

$$N_p = 240(\lg W_p - 0.01)$$

1968 年以后：

$$\frac{1}{a_2} = \frac{\lg 1000 - \lg 100}{840 - 450} = \frac{1}{390}$$

$$a_2 = 390$$

查图得：$b_2 = 8.185$，$\lg b_2 = 0.913$，因此第二直线段方程为：

$$N_p = 390(\lg W_p - 0.913)$$

(3) 计算累积产水量 W_p 和含水率 f_w。

第一直线段取 1959 年：

$$W_p = b \cdot 10^{(N_p/240)} = 1.023 \times 10^{(388.7/240)} = 42.7(\times 10^4 \text{m}^3)$$

$$f_w = \frac{2.3 W_p}{a + 2.3 W_p} = \frac{2.3 \times 42.7}{240 + 2.3 \times 42.7} = 0.29$$

第二直线段取 1975 年：

$$W_p = 10^{(\lg b_2 + N_p/a_2)} = 10^{(0.913 + 751.9/390)} = 693(\times 10^4 \text{m}^3)$$

$$f_w = \frac{2.3 \times 693}{390 + 2.3 \times 693} = 0.80$$

全部计算结果见表 4-3。

(4) 根据后一直线段计算最大累积产油量和最终采收率：

$$N_{p\max} = 390[\lg(21.3 \times 390) - 0.913] = 1172.5(\times 10^4 \text{t})$$

$$R_{\max} = \frac{N_{p\max}}{N} = \frac{1172.5}{2409} = 49\%$$

最终采收率也可根据式(4-31)直接计算：

$$R_{\max} = \frac{390}{2409}\left[\lg\left(21.3 \times \frac{390}{2409}\right) - \lg \frac{8.185}{2409}\right] = 49\%$$

第三节　乙型水驱规律曲线与威布尔预测模型的联解法

一、威布尔模型的建立

Weibull(威布尔)于 1939 年提出的分布模型已成为生命试验和可靠性理论研究的基

础。该模型的分布密度表示为：

$$f(x) = \frac{\alpha}{\beta} x^{\alpha-1} e^{-(x^\alpha/\beta)} \qquad (4-32)$$

式中　$f(x)$——威布尔分布的分布密度函数；

　　　x——分布变量，根据实际问题，分布区间为 $0 \sim \infty$；

　　　α——控制分布形态的形状参数；

　　　β——控制分布峰位和峰值的尺度参数。

若对(4-32)式进行积分，在 x 为 $0 \sim \infty$ 区间内，可以得到 Weibull 的分布函数值等于1，推证如下：

$$F(x) = \int_0^\infty f(x) dx = \int_0^\infty \frac{\alpha}{\beta} x^{\alpha-1} e^{-(x^\alpha/\beta)} dx = -\int_0^\infty e^{-(x^\alpha/\beta)} d(-x^\alpha/\beta) \qquad (4-33)$$
$$= -e^{-(x^\alpha/\beta)} \Big|_0^\infty = 1$$

为将威布尔分布模型用于油气田开发指标的预测，将式(4-32)改写为：

$$Q = \frac{C\alpha}{\beta} t^{\alpha-1} e^{-(t^\alpha/\beta)} \qquad (4-34)$$

式中　Q——油气田的年产量，$10^4 t/a$（油）或 $10^8 m^3/a$（气）；

　　　t——油气田的开发时间，a；

　　　C——由威布尔分布模型转换为油气田开发实用模型的模型转换常数。

油气田的累积产量表达式为：

$$N_p = \int_0^t Q dt \qquad (4-35)$$

将式(4-34)代入式(4-35)并考虑式(4-33)中的变量变换法，t 从 0 到 t 积分得：

$$N_p = C[1 - e^{-(t^\alpha/\beta)}] \qquad (4-36)$$

当 $t \to \infty$ 时，$e^{-(t^\alpha/\beta)} = 0$，则 $N_p = C = N_R$，因此式(4-36)又可改写为：

$$N_p = N_R[1 - e^{-(t^\alpha/\beta)}] \qquad (4-37)$$

在得到上面的结果之后，便可对模型转换常数的性质和作用，作这样的说明：由于威布尔分布模型，在 x 从 0 到 ∞ 区间的分布函数 $F(x) = 1.0$，这相当于实际开发的油气田，在 t 从 0 到 ∞ 区间内的累积产量，即油气田的可采储量。因此，为了能够得到式(4-36)的结果，就必须在式(4-34)中引入模型转换常数 C，该模型转换常数就是油气田的可采储量。因此，可以将式(4-34)再改写为：

$$Q = \frac{N_R \alpha}{\beta} t^{\alpha-1} e^{-(t^\alpha/\beta)} \qquad (4-38)$$

为了确定最高年产量发生的时间，由式(4-38)对时间 t 求导数得：

$$\frac{dQ}{dt} = \frac{N_R \alpha}{\beta} t^{\alpha-2} \left[(\alpha - 1) - \frac{\alpha}{\beta} t^\alpha \right] e^{-(t/\beta)} \tag{4-39}$$

当 $dQ/dt = 0$，必然有 $(\alpha - 1) - \frac{\alpha}{\beta} t^\alpha = 0$ 时，故可以得到最高年产量发生的时间 t_m 为：

$$t_m = \left[\frac{\beta(\alpha - 1)}{\alpha} \right]^{1/\alpha} \tag{4-40}$$

将式(4-40)代入式(4-38)，得到油气田的最高年产量 Q_{max} 的表达式：

$$Q_{max} = N_R \left(\frac{\alpha}{\beta} \right)^{1/\alpha} (\alpha - 1)^{1-1/\alpha} e^{-[(\alpha-1)/\alpha]} \tag{4-41}$$

再将式(4-40)代入式(4-37)，得到油气田最高年产量发生时的累积产量 N_{pm} 为：

$$N_{pm} = N_R \{ 1 - e^{-[(\alpha-1)/\alpha]} \} \tag{4-42}$$

油气田的剩余可采储量(N_{RR})表示为：

$$N_{RR} = N_R - N_p \tag{4-43}$$

将式(4-37)代入式(4-43)得：

$$N_{RR} = N_R e^{-(t/\beta)} \tag{4-44}$$

剩余可采储量的储采比(ω)表示为：

$$\omega = N_{RR}/Q \tag{4-45}$$

将式(4-38)和式(4-44)代入式(4-45)得：

$$\omega = \frac{\beta}{\alpha t^{\alpha-1}} \tag{4-46}$$

剩余可采储量的采油速度为储采比的倒数，故由式(4-46)得到剩余可采储量采油速度(v_o)的表达式：

$$v_o = \frac{\alpha t^{\alpha-1}}{\beta} \tag{4-47}$$

式中 v_o 以小数表示，若改为百分数(%)表示时，式(4-47)改为：

$$v_o = \frac{100 \alpha t^{\alpha-1}}{\beta} \% \tag{4-48}$$

二、威布尔预测模型与乙型水驱曲线的联解法

利用数理统计学中的威布尔分布，研究与推导得到了威布尔预测模型。该模型具有预测油田产量、累积产量和可采储量的功能，其基本关系式分别为：

$$Q_o = at^b \exp\left(-\frac{t^{b+1}}{c} \right) \tag{4-49}$$

$$N_p = \frac{ac}{b+1} \left[1 - \exp\left(-\frac{t^{b+1}}{c} \right) \right] \tag{4-50}$$

$$N_R = \frac{ac}{b+1} \quad (4-51)$$

乙型水驱曲线法首先由我国著名专家童宪章先生以经验公式的形式,于 1978 年提出,基本关系式为:

$$\lg L_p = A + BN_p \quad (4-52)$$

由式(4-52)对时间 t 求导数得:

$$\frac{1}{2.303 L_p} \cdot \frac{dL_p}{dt} = B \frac{dN_p}{dt} \quad (4-53)$$

已知: $\frac{dL_p}{dt} = Q_o + Q_w$; $\frac{dN_p}{dt} = Q_o$; $\frac{Q_w}{Q_o} = R_{wo}$, 故由式(4-53)得:

$$L_p = \frac{1}{2.303B}(1 + R_{wo}) \quad (4-54)$$

将式(4-54)代入式(4-52)得:

$$\lg(1 + R_{ow}) = A + BN_p + \lg 2.303B \quad (4-55)$$

取经济极限水油比 $(R_{wo})_L$, 由式(4-55)得到预测油田可采储量的关系式:

$$N_R = \frac{\lg[1 + (R_{wo})_L] - (A + \lg 2.303B)}{B} \quad (4-56)$$

已知水油比与含水率的关系为:

$$R_{wo} = \frac{f_w}{1 - f_w} \quad (4-57)$$

将式(4-57)代入式(4-55)得:

$$f_w = 1 - 10^{-(A + BN_p + \lg 2.303B)} \quad (4-58)$$

将式(4-50)代入式(4-58)得:

$$f_w = 1 - 10^{-\left(A + B\left\{\frac{ac}{b+1}\left[1 - \exp\left(-\frac{t^{b+1}}{c}\right)\right]\right\} + \lg 2.303B\right)} \quad (4-59)$$

当由式(4-49)和式(4-59)得到预测的产油量和含水率之后,可由下面的公式分别预测油田的产水量和产液量:

$$Q_w = Q_o\left(\frac{f_w}{1 - f_w}\right) \quad (4-60)$$

$$Q_L = Q_o\left(\frac{1}{1 - f_w}\right) \quad (4-61)$$

最高年产量发生的时间 t_m 的计算公式如下:

$$\frac{dQ_o}{dt} = at^{b-1} e^{-\frac{t^{b+1}}{c}}\left(b - \frac{b+1}{c}t^{b+1}\right) = 0$$

即 $b - \frac{b+1}{c}t^{b+1} = 0$ 时,有:

$$t_m = \left(\frac{bc}{b+1}\right)^{\frac{1}{b+1}}$$

最高年产量 $Q_{\max}$ 为：

$$Q_{\max} = a\left(\frac{bc}{b+1}\right)^{\frac{b}{b+1}} e^{-\frac{b}{b+1}}$$

三、模型的求解方法

为了确定预测模型的模型常数 a、b、c 及可采储量 N_R 的数值，对式(4-49)可进行如下处理：

$$\lg \frac{Q_o}{t^b} = \lg a - \frac{1}{2.303c} t^{b+1} \quad (4-62)$$

若设：

$$\alpha = \lg a \quad (4-63)$$

$$\beta = \frac{1}{2.303c} \quad (4-64)$$

则得：

$$\lg \frac{Q_o}{t^b} = \alpha - \beta t^{b+1} \quad (4-65)$$

根据实际的开发数据，首先利用式(4-65)进行线性试差求解，根据最大线性相关系数求出 b，然后利用最小二乘法求得 α 和 β。再由式(4-63)和式(4-64)改写的下式，分别确定模型的常数 a 和 c 的数值：

$$a = 10^{\alpha} \quad (4-66)$$

$$c = \frac{1}{2.303\beta} \quad (4-67)$$

确定出预测模型参数 a、b、c 后，即可根据式(4-51)求解出可采储量 N_R。

在确定预测模型常数 a、b、c、N_R 时，其值是否正确可靠，要利用式(4-49)、式(4-50)、式(4-59)预测的理论产油量、累积产油量和含水率，与实际产油量、累积产油量和含水率进行对比加以确定，而达到最佳拟合效果的参数才是最准确、可靠的。

四、应用实例

[例4-3] 某油田的开发数据见表4-4，试进行分析预测。

表4-4 某油田开发数据

年份	时间(a)	$Q_o(10^4 t/a)$	$Q_w(10^4 t/a)$	$N_p(10^4 t)$	$W_p(10^4 t)$	$L_p(10^4 t)$
1965	1	106.03	7.74	106.03	7.74	113.77
1966	2	63.39	8.52	169.42	16.26	185.68
1967	3	48.42	11.78	217.84	28.04	245.88
1968	4	68.62	1.68	286.46	29.72	316.18
1969	5	81.37	3.86	367.83	33.58	401.41
1970	6	92.08	6.58	459.91	40.16	500.07
1971	7	109.68	13.34	569.59	53.49	623.08
1972	8	123.51	21.89	693.10	75.38	768.48
1973	9	121.73	23.92	814.83	99.30	914.13
1974	10	146.50	36.10	961.33	135.40	1096.73
1975	11	173.16	57.28	1134.49	192.68	1327.17
1976	12	180.14	93.50	1314.63	286.18	1600.81
1977	13	171.21	133.54	1486.35	419.71	1906.06
1978	14	163.19	175.50	1649.54	595.21	2244.75
1979	15	161.12	209.52	1810.66	804.73	2615.39
1980	16	151.50	273.69	1962.16	1078.42	3040.58
1981	17	145.47	347.11	2107.63	1425.53	3533.16
1982	18	130.19	391.83	2237.82	1817.36	4055.18
1983	19	116.26	439.37	2354.08	2256.73	4610.81
1984	20	99.20	436.81	2453.28	2693.59	5146.87
1985	21	89.94	471.18	2543.22	3164.77	5707.99
1986	22	79.87	470.17	2623.12	3634.94	6258.06
1987	23	72.00	491.61	2695.12	4126.55	6821.67

解:(1)根据油气田实际生产数据,进行线性回归,求得乙型水驱曲线的截距、斜率。

将表4-4中的累积产液量(L_p)和相应的累积产油量(N_p)数据,按照式(4-52)的直线关系绘于图4-14,得到了一条很好的直线。经线性回归求得直线的截距$A = 2.60$,直线的斜率$B = 0.000454$,直线的相关系数$r = 0.9997$。

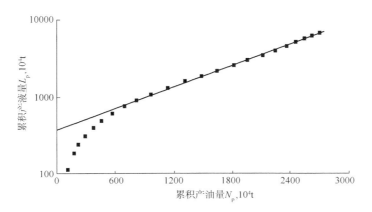

图4-14 该油田的乙型水驱曲线

(2) 确定威布尔预测模型常数 a、b、c。

将表 4-4 内的 Q_o 与 t 的相应的开发数据，按照式（4-65）进行线性试差，根据最大相关系数求出 b（图 4-15），然后求解 a、c，得到：$r=0.999$；$a=19.99$；$b=1.100$；$c=328.10$ 然后，将 a、b、c 的值代入式（4-51）得：$N_R=3178.27 \times 10^4 t$。

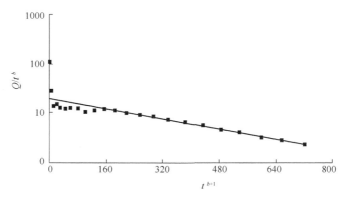

图 4-15 Q_o/t^b 与 t^{b+1} 的半对数关系图

(3) 计算油田年产油量、累积产油量、含水率、可采储量。

将 a、b、c 的数值，分别代入式（4-49）和式（4-50），可以得到预测该油田的理论产油量和累积产油量的相关公式为：

$$Q_o = 19.9 \times t^{1.1} \exp\left(-\frac{t^{1.1+1}}{328.10}\right) \tag{4-68}$$

$$N_p = \frac{19.9 \times 328.10}{1.10+1}\left[1-\exp\left(-\frac{t^{1.1+1}}{328.10}\right)\right] \tag{4-69}$$

再将 A、B、a、b 和 c 的数值代入式（4-59），即得预测该油田含水率的相关公式为：

$$f_w = 1 - 10^{-\left(2.6+0.000454 \times \left\{\frac{19.99 \times 328.10}{1.10+1}\left[1-\exp\left(-\frac{t^{1.1+1}}{328.10}\right)\right]\right\}+\lg(2.303 \times 0.000454)\right)} \tag{4-70}$$

若将上述求得的 A、B 和 N_R 的数值代入式（4-58），可得该油田废弃条件下的极限含水率为：

$$f_w = 1 - 10^{-[2.6+0.000454 \times 3178.27+\lg(2.303 \times 0.000454)]} = 0.913(\text{或} 91.3\%)$$

(4) 计算最高年产量发生的时间、最高年产量。

最高年产量发生的时间 t_m：

$$t_m = \left(\frac{bc}{b+1}\right)^{\frac{1}{b+1}} = 11.6(a)$$

最高年产量 Q_{max} 为：

$$Q_{max} = a\left(\frac{bc}{b+1}\right)^{\frac{b}{b+1}} e^{-\frac{b}{b+1}} = 175.44(\times 10^4 t)$$

(5)绘制油田实际年产量与预测产量对比曲线、实际累积产量与预测累积产量对比曲线。

给定不同的开发时间 t，由式(4-68)、式(4-69)、式(4-70)，可得该油田的预测理论产油量 Q_o、累积产油量 N_p 和含水率 f_w 的数值，列于表4-5，绘于图4-16至图4-18。

表4-5 实际与预测数据对比

时间(a)	$Q_o(10^4 t/a)$		$N_p(10^4 t)$		$f_w(\%)$	
	实际	预测	实际	预测	实际	预测
1	106.03	19.93	106.03	9.50	6.8	0.00
2	63.39	42.29	169.42	40.54	11.8	0.00
3	48.42	64.92	217.84	94.17	19.6	0.00
4	68.62	86.85	286.46	170.14	2.4	0.00
5	81.37	107.35	367.83	267.39	4.5	0.00
6	92.08	125.83	459.91	384.17	6.7	0.00
7	109.68	141.78	569.59	518.20	7.3	0.00
8	123.51	154.85	693.10	666.77	15.1	0.00
9	121.73	164.78	814.83	826.86	16.4	0.00
10	146.50	171.46	961.33	995.26	19.8	15.12
11	173.16	174.90	1134.49	1168.70	24.9	29.20
12	180.14	175.20	1314.63	1344.00	34.2	41.05
13	171.21	172.60	1486.35	1518.14	43.7	50.86
14	163.19	167.40	1649.54	1688.34	51.8	58.87
15	161.12	159.98	1810.66	1852.20	56.5	65.35
16	151.50	150.74	1962.16	2007.69	64.4	70.54
17	145.47	140.11	2107.63	2153.21	70.5	74.70
18	130.19	128.53	2237.82	2287.59	75.1	78.02
19	116.26	116.41	2354.08	2410.10	80.0	80.66
20	99.20	104.12	2453.28	2520.36	83.1	82.77
21	89.94	92.00	2543.22	2618.40	84.0	84.44
22	79.87	80.31	2623.12	2704.50	85.5	85.78
23	72.00	69.28	2695.12	2779.24	87.2	86.85
24	—	59.07	—	2843.34	—	87.70
25	—	49.79	—	2897.69	—	88.38
26	—	41.48	—	2943.24	—	88.92
27	—	34.17	—	2980.99	—	89.35
28	—	27.83	—	3011.91	—	89.69
29	—	22.42	—	3036.96	—	89.96
30	—	17.85	—	3057.03	—	90.17

续表

时间(a)	$Q_o(10^4 t/a)$ 实际	$Q_o(10^4 t/a)$ 预测	$N_p(10^4 t)$ 实际	$N_p(10^4 t)$ 预测	$f_w(\%)$ 实际	$f_w(\%)$ 预测
31	—	14.06	—	3072.92	—	90.33
32	—	10.95	—	3085.38	—	90.45
33	—	8.44	—	3095.03	—	90.55
34	—	6.43	—	3102.43	—	90.62
35	—	4.85	—	3108.04	—	90.68
36	—	3.61	—	3112.24	—	90.72
37	—	2.66	—	3115.36	—	90.75
38	—	1.94	—	3117.64	—	90.77
39	—	1.40	—	3119.30	—	90.79
40	—	1.00	—	3120.50	—	90.80
41	—	0.71	—	3121.34	—	90.80
42	—	0.49	—	3121.94	—	90.81
43	—	0.34	—	3122.35	—	90.81
44	—	0.23	—	3122.63	—	90.82
45	—	0.16	—	3122.83	—	90.82
46	—	0.11	—	3122.96	—	90.82
47	—	0.07	—	3123.04	—	90.82
48	—	0.05	—	3123.10	—	90.82
49	—	0.03	—	3123.14	—	90.82
50	—	0.02	—	3123.16	—	90.82

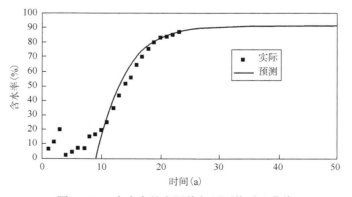

图 4-16 含水率的实际值与预测值对比曲线

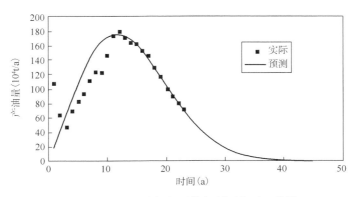

图 4-17　产油量的实际值与预测值对比曲线

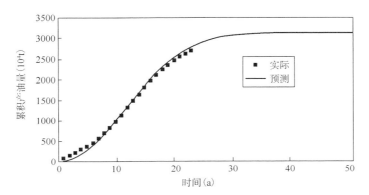

图 4-18　累积产油量的实际值与预测值对比曲线

第五章 油田开发调整理论与技术

油田开发方案实施以后,是一个不断调整变化的进程。随着油田开发的进行,稳产条件及开采条件逐渐发生变化,对油层的认识持续丰富和加深,这就有必要进行开发部署的调整。

第一节 油田开发信息监测与分析评价

一、油田开发信息监测

所谓油田开发信息监测,就是对油田开发过程中的各种参数进行实时记录,目的是为开发分析评价提供数据参数。油田开发信息监测的主要内容有油气水产量、地层压力、产油(液)剖面与吸水剖面、产出流体性质、油气水界面运动规律、水驱前缘推进规律、地层含水饱和度变化规律、出砂规律、井间连通性等,并把监测结果整理成一系列的图表和曲线,以便进行分析、对比和评价。地层压力通过定期的压力恢复试井进行监测。

油井产液或产油剖面的监测需通过定期的生产测井才能完成。生产测井的种类很多,如流量计、电阻率、放射性同位素和噪声等,生产测井的解释结果可以做成图 5 – 1 的形式。图中曲线显示油井有两个产层,上产层为主力油层。一年前的产液剖面显示,上产层产油,下产层已 50% 产水。一年后的产液剖面显示,上产层已开始产水,下产层已 100% 水淹,应立即采取措施对下产层进行彻底封堵。

注水井吸水剖面的监测与采油井产液剖面的监测十分类似,可根据吸水剖面的监测结果制定相应的调剖和堵水措施。

地层含水饱和度的监测可以通过定期的井点或井间示踪剂监测完成,也可以通过钻专门的取心检查井来完成。井间示踪剂监测可同时监测油水井之间的油层连通性。

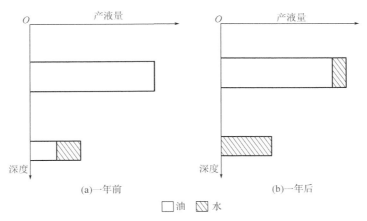

图 5-1 油井产液剖面

油气水界面或水驱前缘移动规律的监测,需要专门的观察井才能完成。通过观察井定期测量油水界面和水驱前缘的位置,即可完成油水界面和水驱前缘移动规律的监测(图 5-2、图 5-3)。同时,还可以用数值模拟方法进行跟踪研究。

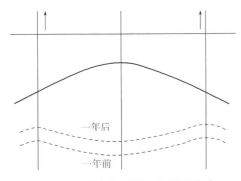

图 5-2 观察井观测油水界面移动

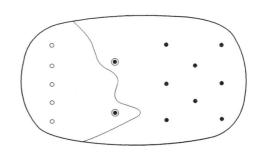

图 5-3 观察井观测水驱前缘移动

油田开发信息监测的内容十分广泛,可以根据油气田开发的具体需要,进行监测内容和监测方案的设计。监测结果直接反映油藏的地层特征和开发措施的正确与否。

二、油田开发分析评价

所谓油田开发分析评价,就是采用一定的方法和手段,研究整理分析评价油田开发过程中获得的各种动态数据资料,逐步提高对油气藏地质特征和油气运动规律的认识程度,以便不断修正地质模型和调整油田开发方案,使油田开发工作不断得到改进并在严密的控制之下良性运行。

1. 地质特征再认识

在油气藏评价阶段,油气藏地质模型的建立是通过大量间接的资料和十分有限的钻井及试采资料进行的,因而对油气藏地质特征的认识是不够全面的。随着开发过程的不断进行,又将增添许多有关油气藏的静态和动态资料。为使油气藏地质模型更加接近真实情况,有必要对油气藏地质特征进行再研究,修正以往由于资料有限而得出的错误认识,补充以往认识上的空缺部分。地质特征再认识的方法和内容与油气藏评价阶段的工作基本相同,只是地质特征再认识是一项反复进行永不间断的研究工作。

2. 储量及可采储量核算

由于新资料信息的增加,对油气藏地质特征的认识会有所改变,有时甚至是较大的改变,油气藏地质模型也将得到进一步修正。由于储量及可采储量规模是决定采用何种开发原则的物质基础,因此有必要在油气藏地质模型修正之后进行储量复算及可采储量的进一步核实。由于在开发过程中不断对油气藏地质模型进行修正,因此储量及可采储量的复算工作也必须反复进行。由于开发过程中的动态资料不断增加,储量计算工作除采用常规的容积法之外,常常辅之以一些动态方法,如物质平衡方法、水驱曲线方法和数值模拟方法等。

3. 储量动用状况分析

一定的开发系统不可能动用所有的地质储量,油藏工程师的任务之一就是采用有效手段和措施动用尽可能多的地质储量。储量动用程度越高,采出的油气也就越多。因此,在开发过程中,分析油气储量的动用状况,并研究提高储量动用程度的措施显得十分重要。

矿场上常用吸水和产液剖面测试资料、密闭取心资料、分层测试和单层生产资料等分析研究注入剂垂向波及状况、水淹水洗状况和油层出油出液状况;应用矿场测试资料和油藏工程分析方法研究注入剂的波及状况及驱油效率;应用水驱曲线分析储量的动用状况及变化趋势;应用物质平衡和数值模拟方法分析油藏的剩余油分布状况等。

4. 产量构成分析

油气田开发工作的核心是油气产量,油气产量的高低及其构成反映了油气田开发水平和开发所处的阶段,因而油气产量就成了油气田动态分析的主要对象之一。通常将油气产量变化分成三组曲线:(1)产量构成曲线,包括老井日产水平、措施井日产水平、新井日产水平;(2)生产综合曲线,包括日产油气水平、日产液水平、日注入量水平、综合含水、综合气油比等;(3)递减曲线,包括自然递减曲线和综合递减曲线。

通过对以上三组曲线的分析研究,可以了解油气田目前的开发状况,也可以预测油气田未来的开发动态,并能提出减缓递减和稳定油气产量的措施及建议。

5. 注采平衡分析

注采平衡是保持油藏能量和实现长期高产稳产的重要保障,开发过程中需及时分析掌握油藏的注采平衡状况。注采平衡分析主要研究以下内容:

(1)注采比的变化与压力保持水平的关系、压力系统和注采井数比的合理性;

(2)合理的压力保持水平,能量利用是否合理,油藏亏空状况,配产、配注方案的效果;

(3)不同开发阶段的合理压力剖面、注水压差和采油压差。

6. 增产措施和油层改造

为合理开发油气资源和充分利用驱动能量,开发过程中经常进行增产、增注和改造油气层的各种措施。在每次增产措施实施之前,都要进行措施的可行性分析;措施实施之后,要进行措施的效果分析、增产增注量和有效期的分析,并分析对油气田稳产和控制递减的影响。

7. 剩余油分析

剩余油是指油田生产到某个时间点为止,仍然剩余在地下未被采出的原油。剩余油的成因机制主要包括三个方面:(1)未波及;(2)波及不充分;(3)洗油效率不够。剩余油是油田工作的核心,涉及两个关键环节,分别为剩余油描述和剩余油挖潜。剩余油描述又是剩余油挖潜的基础。

剩余油描述的内容包括:(1)剩余油的位置;(2)剩余油的量,如储量、丰度、单储系数、剩余油饱和度等;(3)剩余油的约束机制,即剩余油的形成机理。

8. 油田开发过程分析

油田开发过程分析是一项极其庞杂的任务,它绝不是仅用于某个单一指标的分析或孤立地分析所有指标这样的工作,它需把油田开发的方方面面综合起来进行考虑,形成一个总的指导开发过程的理念体系。一般说来,油田开发过程分析需进行如下几项工作:

(1)建立油气田开发数据库:记录油气田静、动态资料,并为充分享用油气田数据资源提供便利。

(2)分析注采系统的适应性:分析方案设计的注采系统是否适应实际需要,是否适应开发过程中高、中、低各个含水阶段,是否适应于自喷、抽油、气举等各种采油方式,是否需要调整注采系统,驱油能量是否充足等。

(3)分析油气层潜力:分析储量动用程度高低,剩余油分布状况,驱替方式和工作制度是否得当,新的提高可采储量和采出程度的措施。

(4)分析三大规律:产量递减规律、含水上升规律、压力变化规律。通过对这三大规律的分析,对油藏生产有一个宏观把握,然后,根据三大规律反映出的问题,提出相应的应对措施。

(5)增产措施和开发调整效果分析:分析开发层系、开发井网、注采系统的调整,开发方式的调整,配产、配注的调整,以及压裂、酸化等增产措施调整,是否有效,有效程度如何。

(6)经济效益分析：分析所采用的开发系统其经济效益如何，采油成本及投资收益率等经济指标是否合理，开发系统是否需进行大的调整等。

第二节　油田开发调整的内容

根据开发监测和开发分析的结果，对原设计开发方案进行综合评价。一般说来，早期根据较少信息设计的开发方案，多少都有一些不适应油藏实际情况的地方。为提高油藏的开发效果，需对原设计开发方案进行适当的调整，就是按照新的地质认识和当前的经济技术条件，重新编制一套开发方案，称作调整方案。调整方案的编制方法与开发方案的编制方法基本相同。

油田开发调整的内容包括：(1)开发层系调整；(2)开发井网调整，涉及井网密度、注采系统和井网形状等；(3)开发方式调整；(4)采油工艺措施调整。

开发调整可以是全面的，也可以是局部的，多数情况下只进行局部的调整。

开发层系调整，一般是在新的地质认识基础上，进行层系的细化。原开发层系，在经过一定时间的生产之后，暴露出了许多层间矛盾。高渗透层被严重水淹之后，低渗透层中的原油却尚未动用或动用程度较低而成为剩余油。此时，为提高原油的采出程度，应将开发层系进一步细化(图5-4)。开发层系调整一般坚持"先粗后细"的原则。

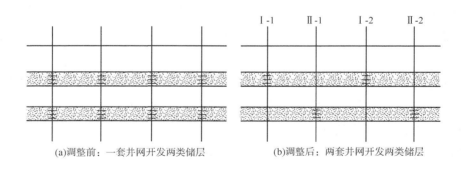

(a)调整前：一套井网开发两类储层　　　　(b)调整后：两套井网开发两类储层

图5-4　开发层系的细化

开发井网调整，一般是在新的地质认识基础上，进行井网的加密和注采井网系统的调整。原开发井网，在经过一定时间的生产之后，表现出了一定的不适应性，主要表现为井距太大，对砂体或储量的控制程度较低。此时，为提高原油的采出程度，应将开发井网进一步加密(图5-5)，为了提高油田产量，井网形式的调整，一般从高注采井数比的井网向低注采井数比的井网调整(图5-6)。井网调整一般坚持"先稀后密"的原则。

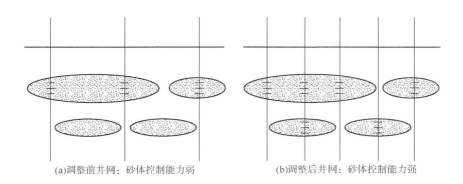

(a)调整前井网：砂体控制能力弱　　　　(b)调整后井网：砂体控制能力强

图 5-5　开发井网的加密

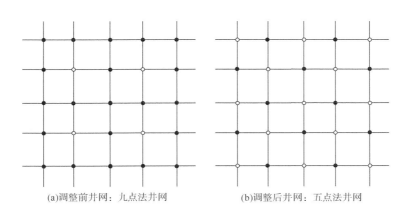

(a)调整前井网：九点法井网　　　　(b)调整后井网：五点法井网

图 5-6　开发井网形式的调整

● 注水井　　○ 生产井

开发方式调整,一般是在新的地质认识基础上,从衰竭式开采向注水(气)开发调整,或从注水开发向 EOR 方法调整。基本原则是从低驱油效率的驱油方式向高驱油效率的驱油方式进行调整。

小的开发调整随时都在进行,如工作制度的调整、液流方向的调整、射孔层位的调整,以及开采剩余油的局部井点加密、油井转注、调剖、堵水和压裂等 EOR 方法;大的开发调整一般都有 3~5 年以上的时间间隔,如开发层系的调整、开发方式的调整、开发井网的调整等。因此,开发调整是一项永无止境的工作。

开发调整一般是根据油气生产的需要而进行设计的,以能够增加油气产量和提高开采效益为原则。

开发调整的根本目的就是通过各种开发措施改变地下的渗流场,让不流动的油流动起来,让流动慢的油流动得快些;逐渐扩大注入水的波及范围,阻断注入水的优势渗流通道,提高注入水的驱油效率。

第三节　喇萨杏油田开发调整经验与展望

油藏多样性、复杂性、隐蔽性和时变性决定了油田开发必然要经历多次开发调整的过程,每次开发调整都应该建立符合本油田、本开发阶段实际情况的开发模式。由于多次采油并存、各种开发方式同在、油层封堵等情况,喇萨杏油田开发调整和建立开发模式更为复杂。

一、5 种开发模式

自 1960 年投入开发以来,喇萨杏油田经历了基础井网、一次加密、二次加密、三次加密、开采方式调整、注采系统调整、注采结构调整、注水细分调整、聚合物驱等开发调整和提高采收率技术工业化推广阶段,形成了以下 5 种具有代表意义的开发模式。

1. "早期内部横切割行列注水保持地层压力"开发模式

油田开发初期,条件艰苦,但仍然开展了著名的"十大开发试验"。在试验研究、战略研究并借鉴国内油田开发经验基础上,结合喇萨杏油田具体特征,开创性地提出了"早期内部横切割行列注水保持地层压力"开发模式。

(1)早期注水。以注水为纲,保持压力开发,使油井长期保持旺盛的生产能力。注采比保持 1 左右,地层压力保持在原始地层压力附近。

(2)行列井网。采用大切割距行列井网开发,两排水井之间布三排油井,第一排油井与注水井排距离是 600m,油井井排距离是 500m,井距均为 500m。后来投产的开发区对上述井排距有所调整。

(3)油层性质最好的葡一组油层独立一套开发层系,其他油层一套开发层系。在之后执行过程中各开发区有所调整。

(4)喇嘛甸油田和高台子油层作为后备储量先不动用,伺机开发。

(5)保持地层压力自喷开采,生产井井底压力允许略低于饱和压力 0.5~1 MPa。

(6)在合理划分开发层系基础上,力争做到在注水井内分层分配吸水量,控制水线均匀推进。

(7)先投产靠近注水井两侧生产井,中间留下一个"小仓库",待第一排生产井 5 年见水后,再将中间井排投产,保持开发区 3% 左右的采油速度稳产 10 年。

(8)争取多采无水原油,5 年内基本不见水,但要充分准备长期开采含水原油。

(9)大力加强选择性注水、选择性堵水和选择性压裂技术研究。

该开发模式具有开创性,有效指导了喇萨杏油田全面开发。从 1960 年开始投产,年产油量一路飙升,到 1976 年油田全面投入开发,年产油量达到 5000×10^4t 以上。

2. "储量分步动用、接替稳产"开发模式

1976年喇萨杏油田基本全面投入开发,决策层提出了"高产上五千,稳产再十年"的目标。针对这个目标,争论很大,因为世界上没有哪个油田全面投产后能够稳产10年。为此,开展了深入而广泛的研究工作,建立了"储量分步动用、接替稳产"开发模式。

(1)依靠"四个立足",实现"五五"期间稳产。深化以分层注水为核心的"六分四清"综合调整,立足基础井网、立足主力油层、立足"六分四清"、立足自喷开采。

(2)实施"三个转变",即主要措施由"六分四清"综合调整转变为井网加密、主要调整对象由主力油层转变为非主力油层和薄差油层、开采方式由自喷转变为机械采油,进行以细分开发层系为主的井网一次加密调整。

(3)调整注采系统,提高储量控制程度,保持和恢复地层压力。由于油水井数比大(3.5∶1),油井供液受效方向少,平面注水调整困难,逐步调整到2∶1以内,开发效果得到明显改善。

(4)开展二次井网加密调整,提高薄差油层和表外储层动用程度。一次井网加密调整以后,仍有发育规模小、物性差的油层难以动用,且原来没有计算储量的物性相对较好的表外储层也具有出油能力,以这些油层为调整对象实施二次加密调整。

(5)优选二次加密调整后仍具有调整潜力的区块和油层实施三次加密。研究表明,二次加密后仍有部分油层动用差或不动用,加上表外储层潜力,大部分地区具有继续进行加密调整的物质基础,可实施三次加密调整。

该开发模式建立了陆相油田分层注水开发理论和技术,确立了大庆油田开发水平在世界的领先地位,支撑了大庆油田长期高产稳产,创造了巨大的经济和社会效益。

3. "结构调整、稳油控水"开发模式

进入20世纪90年代,喇萨杏油田综合含水已超过80%,进入高含水后期。以细分层系为主的一次加密调整已经结束,开始进行以调整薄差层和表外储层为主的二次加密调整,年产油量达到5600×10^4t这个产量高点。从油田整体开发潜力状况来看,二次加密和外围油田具有较大建产能力。面对产液量快速上升,各开发区、各类井和各类油层开发状况差异较大的状况,提出了"结构调整、稳油控水"开发模式。

(1)根据各类油井、各类油层、各类区块生产状况进行分类调整。

(2)控制含水高的油层、油井、区块,放开含水低的油层、油井和区块,改善动用差和未动用的油层开发效果。

(3)应用分注、压裂、井网加密、注采系统调整和压力系统调整等常规技术手段,调整注水结构、产液结构、储采结构。

(4)针对问题明显、开发状况差异大的油井、油层、区块集中治理,总结经验全面推广并

使之常态化。

(5)调整目标是控制含水上升,控制产液量快速增长,保持油田稳产。

"结构调整、稳油控水"综合调整技术是继"分层注水开发""细分层系井网加密调整"之后的又一次技术飞跃,是大庆喇萨杏油田水驱开发技术的丰富和发展。为油田生产日常管理、综合调整指明了方向,使多油层砂岩油田注水开发理论、思想和技术体系更加完善。

4."五四三二一"精细开发调整模式

经过多次开发调整后,剩余油和开发调整潜力高度分散,为此研究形成了"五四三二一"精细开发调整模式,即"五个不等于""四个精细""三个层次""两个控制""一条特高含水期精细高效开发的新路子"。

(1)以"五个不等于"(油田高含水不等于每口井都高含水,油井高含水不等于每个层都高含水,油层高含水不等于每个部位、每个方向都高含水,地质工作精细不等于认清了地下所有潜力,开发调整精细不等于每个区块、井和层都已调整到位)的潜力观认识油田开发潜力,精细量化注水标准,采用传统技术油水井多项措施协同挖潜。

(2)针对特高含水期油田开发矛盾加剧的状况,确立"四个精细"(精细油藏描述、精细注采系统调整、精细注采结构调整、精细生产管理)的挖潜思路,实施"三个层次"(平面结构调整、层间结构调整、层内结构调整)立体调整,减缓"平面、层间和层内"三大矛盾,实现"两个控制"(控制产量递减、控制含水上升)的目标。

(3)常规技术,常用常新。分层注水、压裂、堵水、换泵、完善单砂体注采关系、井网加密、注采系统调整和压力系统调整等常规技术向精细化发展。

(4)以示范区为先导,总结经验,全面推广。大庆油田从2008年提出"四个精细"以来,含水上升率控制在0.5%以内,产量自然递减率和综合递减率分别控制在9%、6%以内,取得较好调整效果,使大庆油田5000×10^4t以上稳产27年之后又在4000×10^4t以上稳产14年。

5."聚驱水驱并行"开发模式

1996年,聚合物驱全面推广应用使喇萨杏油田开发工作形成了"聚驱水驱并行"的局面。一方面聚合物驱技术不断推广,水驱油层不断转入聚合物驱;一方面水驱调整按照"五四三二一"开发模式不断精细化。因此说,这一时期水驱和聚驱两种开发模式基本处于并行状态,是水驱开发和化学驱提高采收率技术全面结合的过渡阶段。

(1)优先选择聚驱技术成熟、效果好的一类油层开展聚驱三次采油,之后进一步优化技术参数和技术配套,向二类油层推广。

(2)一套开发层系控制在8~10m,以聚驱三次采油层系井网为主,目的层封堵水驱。为了提高配制、注入和井网等设备利用率,纵向上只安排一套层系聚驱,逐级上返。

(3) 根据聚合物体系黏度大、油层吸入能力低等特点,采用五点法面积井网,聚驱控制程度达到70%以上。

(4) 根据油层性质优化聚合物分子量、溶液体系浓度和注入段塞,提高聚合物和油层配伍性,保证体系黏度不低于40 mPa·s,实现分层注入,研究探索分层分质注入方式的可能性。

1996年,聚驱全面工业化推广应用,在水驱基础上提高采收率10个百分点以上,长期保持1000×10^4t以上生产规模,为大庆油田5000×10^4t和4000×10^4t稳产作出了巨大贡献。

喇萨杏油田成为世界最大的化学驱研究和生产基地,其聚驱技术的推广应用是世界油田开发领域的重大事件,引领了提高采收率技术的发展方向,使喇萨杏油田成为分层注水开发技术和三次采油技术研究与应用最成功的典范。

二、开发模式演变的内在规律性

应用事物发展变化的辩证规律分析喇萨杏油田开发模式的演变,可以看出其内在联系,为今后的研究工作提供思想准备。

(1) 油藏具有多样性、复杂性、隐蔽性和时变性,这些特性决定了油藏的认识过程和开发调整过程不可能一次完成。每次的调整过程录取了大量的新资料,同时又有了新的动态反应,这些为深化认识创造了条件。因此,油田开发调整必然要经历多次认识、多次调整、循环往复的过程。

(2) 喇萨杏油田前4种开发模式的演变过程是认识油田、改造开采条件不断深化的过程,也是开发调整对象由粗到细的过程。第1个开发模式的建立具有开创性意义,核心是"开发一个大油田"。其开发模式的基本内容是从整体上、全局上考虑问题。第2个开发模式是在油田投入开发以后,在相同开采条件下好油层与差油层之间动用状况差别巨大情况下建立的。好油层继续用基础井网开发,薄差油层则采取细分层系和井网加密开采。因此,该开发模式的核心是"两类油层区别对待,分步动用,接替稳产"。第3个开发模式是在各套井网并存、各类油层开发状况存在较大结构性差异情况下建立的,核心是"结构调整""三分一优""后进赶先进",调整优化开发状态。第4个开发模式是在油田经过上述3个开发模式的开发调整以后,水驱剩余油潜力分布零散情况下建立的,核心是"精细挖潜、吃干榨尽"。

(3) 各种提高采收率技术的推广应用把油田开发主要调整方向由扩大波及体积转向提高驱油效率和治理无效循环。传统水驱开发调整似乎到了"尽头"。研究这4个开发模式,尽管差别很大,但共同点是调整方向相同,都是"扩大波及体积、挖掘剩余油"。提高采收率除了这个方向以外还有两个方向:一是提高驱油效率,二是调整驱替程度差异。关于聚驱是以扩大波及系数为主还是以提高驱油效率为主还有争论,但是二元、三元、泡沫复合驱可以提高驱油效率是没有疑义的,同时也可以扩大波及体积。在无效循环出现之前驱替程度差异表现为先进和后进的矛盾,进入特高含水期后出现无效循环,驱替程度差异的矛盾性质发生了变化,成为制约开发效果的主要影响因素。无效循环是一个"毒瘤",只有破坏作用,没有任何好处,必须治理。因此,今后建立开发模式的调整方向应转移到提高驱油效率、治理无效循环、缩小驱替程度差异上来,从而开始新的开发模式的演变过程。

(4)开发模式的演变严格践行了"实践、认识、再实践、再认识"循环往复、循环上升的哲学规律。水驱开发调整及其建立的开发模式经历了由粗到细和几次"实践、认识、再实践、再认识"的过程,然后技术升级进行三次采油,开始了新的开发调整"实践、认识"的循环过程。从油田开发专业特点和过去的实践经验来看,每次"实践、认识"的循环,技术点只有增多没有减少,技术难度只有增大没有减小,管理点只有增多没有减少。原来行之有效的做法不能放弃,新技术、新做法不断增加。

(5)新的开发阶段问题会更多、情况会更加复杂。今后必将逐步进入水驱、聚驱、复合驱、气驱、聚驱后、微生物驱等"多元并存",无效循环治理、表外开发、低效井治理、多种方式提高采收率"多措并举"的开发阶段,二次、三次、四次采油甚至五次采油紧密交织在一起,各套层系、井网交织在一起。研究和开发调整的难度会更大,需要投入更多的研究力量和开发成本。

综上所述,油田开发调整挖潜的过程总是"先大后小、先肥后瘦、先粗后细、先易后难"。"先大后小、先肥后瘦"符合解决问题抓"主要矛盾"的思想,"先粗后细、先易后难"符合技术发展规律。事物在某一方向发展到极致就会出现新的转机,开辟新的天地,这完全符合"实践、认识、再实践、再认识"的哲学规律。

三、今后喇萨杏油田开发模式探讨

新的开发阶段情况复杂、问题复杂,需要从急需解决的问题入手开展研究。

1. *存在问题*

(1)无效循环已经成为特高含水期水驱开发的主要矛盾,消耗成本、抑制低渗、稀释有效,是平面、层间、层内三大矛盾加剧的根本原因。治理低效无效循环必将成为今后水驱开发调整的主要方向。两类油层开发矛盾加大体现在四个方面:一是各种调整措施实施以后剩余未动用油层与已动用之间的矛盾加大;二是已动用油层中好油层(主要是无效循环层)和差油层之间的矛盾加大;三是表外储层和表内储层之间物性差异大,矛盾也大;四是聚驱后低渗透油层中滞留聚合物更多,因此两类油层矛盾加大。

(2)油田已全面进入特高含水开发,各类井含水逐渐接近,平均单井产量下降到 2 t/d 以下,低效井数日益增多。低效井的治理与综合利用成为今后必须面对的问题。

(3)以开采表外储层为主的井网加密调整效果越来越差,与表内储层合采开发矛盾大,含水高、产量低、低效井比例大。需要采取更加强化的技术措施才能减缓开发矛盾,因此应优选有利地区独立开发层系或有针对性地强化注水。

(4)当前三次采油是主体技术,必须独立建立层系井网,逐级上返。水驱则处于配合地位,油层逐步转移到三采后,水驱井网注采关系被肢解和蚕食,必须研究该套井网的利用问题。

(5)聚驱后进一步提高采收率技术成熟,推广应用时间是影响全局的不确定性因素。这种不确定性把技术发展和推广进程加入层系井网调整中来,层系井网调整不再是单纯的技术问题。

(6)综合油田动态和取心井资料分析发现,水驱剩余可动油可分为五大类:第一类是未见水油层剩余油,占比较少。说明现有井网平面控制程度很高,开发效果好。第二类是表外储层剩余可动油,占比中等,是水开发调整的主要对象。第三类是见水油层中未水洗部位可动油,这部分剩余油最多,是三次采油的主要挖潜对象。见水油层中已水洗部位中的剩余可动油可分为两部分,其中一部分就是第四类剩余油——水洗油层中不能有效采出的剩余可动油,因已动用而被忽视,是控制无效循环的主要挖潜对象;另一部分就是第五类剩余油:剩余可采储量,现有井网可有效采出。

2.建立开发模式的基本原则

由于今后的开发调整二次、三次、四次采油并存,表外挖潜、无效循环治理、低效井治理同在,聚驱后推广时间不确定等复杂因素,且不同地区油层状况、开发状态、潜力状况也存在很大差别,难以建立统一的调整模式,但也有很多共性问题可以作为原则性指导意见。

(1)三次采油井网、层系上返先规划、先定位。

(2)表外开发要作为重点水驱调整对象,优选有利地区独立开发,全面强化注水调整。

(3)聚驱后技术推广时需设定不同的设想预案,研究层系井网组合方式。

(4)表内水驱层系井网分段组合,与无效循环治理、低效井治理和三采层系变化相结合。

(5)要减缓加剧的开发矛盾,就需要实施比原来更加强化的技术措施。

3.需要重点研究的问题

要实现上述设想,建立新的开发模式,必须强化以下五个方面的研究:一是分类细化调整潜力研究,二是加快聚驱后主体技术研究,三是表外储层有效开发方式研究,四是无效循环治理方式和技术研究,五是层系井网组合方式研究。

四、认识与建议

(1)由于油藏的复杂性、多样性、隐蔽性和时变性,油田开发必然要经历多次认识、多次调整、循环递进的开发过程,开发模式本地化是油田高效开发的基本技术路线,是油藏工程研究的重要组成部分。因此必须重视全过程开发战略研究,对油田开发过程进行预判,超前进行技术储备。

(2)喇萨杏油田水驱开发模式演变完成了"扩大波及体积、挖掘剩余油"的由粗到细的开发调整过程,是"实践、认识、再实践、再认识"哲学规律的完美体现。聚驱规模不断扩大,开启了开发方式升级后新的开发调整阶段。

(3)根据油田开发状况,就如何建立今后的油田开发模式取得了三点认识:一是水驱开发调整将转变到"治理无效循环、调整驱替程度差异、挖掘极端分散剩余油"方向上来,二是新开发模式必然要采取"多元并存、多措并举"的技术措施,三是聚驱后等待时间的不确定性对新开发模式的建立影响较大。

(4)注水无效循环是特高含水期油田开发的主要矛盾,对油田开发产生了三大影响,即"消耗成本、抑制低渗、稀释有效"。治理无效循环,调整驱替程度差异,具有提高采收率近5%的潜力。

(5)表外储层开发潜力大,与有效层合采层间矛盾大,应强化开发方式的研究。

(6)要充分认识今后油田开发的复杂性,加大理论研究、战略研究和技术研究的力度,按照实践和认识的基本规律,研究建立优化的开发模式,不断推进油田开发理论和技术进步。

第四节　关于变革性开采技术的探讨

油田开发是一个系统工程,只有从历史观、大局观、整体观出发,运用辩证和发展的哲学思维,进行分析、设计和跟踪调整,才能取得最佳的技术、经济和社会效果。

一、缩小尺度是大幅度提高石油产量和采收率的技术关键

一项技术必须具备两个特征,才可以称为"革命性"或者"变革性"技术。众所周知,第一次工业革命是蒸汽机,第二次工业革命是电,第三次工业革命是计算机。这三大技术的共同点:一是带来显著的变化,有电灯电话和没有的时候比较,变化是显著的、革命性的;二是实现规模化的应用,家家户户都用到电、用到计算机,可谓革命性的规模化应用。

在石油行业历史上,聚合物驱油技术可以称得上是行业性革命性技术,一是效果显著,矿场应用提高采收率10%;二是实现了石油开采行业内的规模化应用。

石油开采行业下一次行业革命的技术关键就在于尺度的缩小。从勘探手段、地质研究、油藏描述,到信息监测、静动态分析、地质模型和油藏数值模拟,再到相关的工具、工艺和工

程,尺度逐步缩小,是关键突破点,必将带来革命性的产量和采收率突破。

二、"治未病"事半功倍

对于一个目标油田而言,在未对它进行建设开发之前,我们就清晰地知道,这个油田或早或晚会形成注水低效无效循环通道,能量的低效无效循环问题成为制约油田开发提效的关键。所以,从一开始的时候,就应该将低效无效循环问题纳入关注重点之一,每一年都针对低效无效循环通道的发展情况、水驱油前缘推进位置、能量的效率等进行跟踪、分析和评价,并且及时地制定应对措施策略,实时地加以人为干预,对低效无效通道的发展加以控制,提高每一年的能量效率。这样可以起到少花钱、效果佳的良好作用,油田开发效果可以达到最佳程度。

三、坚持以水驱为主体框架

注水开发具有成本低、技术经济效果好等优势。坚持在水驱为主的大框架下,通过创新各种工艺措施,扩大波及体积和已波及区波及倍数,不断提高水驱采收率。在确认水驱技术达到极限之后,再考虑实施聚合物驱等化学驱技术。

四、发展"颠覆性"思维方式,创新变革性技术

摒弃"贴地皮"思维模式,"颠覆性"地琢磨,创新研发变革性技术,包括原始性创新和组合式创新。基于"不着边际"思维的变革性创新,能够引发翻天覆地的改变,大幅度提高石油采收率。只有想得到,才能做得到。下面提出几个"颠覆性"的技术蓝图。

1. 小威力液体炸药"微爆炸"技术

发明创新能够易于注入地层孔隙网络、能够依靠局部憋压等原理起爆、爆炸强度相当于过年时候小孩子们"呲花"那样的微爆炸。这种微爆炸技术不仅可以改变流体分布、启动剩余油,而且可以改变孔喉网络的固体结构、改变流动通道,本质上是不同于传统的提高采收率技术。

2. 纳米机器人技术

基于现代机器人技术,创新研发能够在地层孔隙网络中不断前进的纳米机器人,实现大幅度提高石油采收率。这种纳米机器人可以是前端带有螺旋桨样式的旋进装置,该装置依靠流体压力差驱动,从而带动整个机器人向前运动,在运动过程中,激发流体旋转、启动剩余油和驱动流体向前流动。

3. 油水井"互换"技术

将油井转为水井、水井转为油井,彻底改变液流方向。可能在调整之初,产出液全是水,

但是经过一段时间排水以后,必定见效。油水井互换角色,从渗流机理上来讲,是有道理、立得住脚的。在生产中,需要大刀阔斧地下定决心,首先以小区块先导性试验开始,逐步总结经验和扩大规模。

五、管理很重要

任何工程,要取得好的技术、经济和社会效果,都离不开"落地""落细""落实"的管理工作。例如,对油藏进行"网格化"管理。将地下划分成尽可能小尺寸的几万、几十万、几百万甚至上千万的网格,工程管理(包括静动态数据、测试资料、分析评价等)都是"对着网格说话",而不是"对着井、层、方向说话",这是尺度上的缩小,更是管理上的升级,这样的管理必然有力支撑大数据、智能化、智慧化等发展,提升油田开发效果。在谋划的时候,大家畅所欲言,各抒己见;在敲定以后,大家坚决地无条件执行,抓好落地;在过程中,要做好监督监管,促进落实。

参考文献 References

[1] 石成方,吴晓慧. 喇、萨、杏油田开发模式及其演变趋势[J]. 大庆石油地质与开发,2019,38(5):45-50.

[2] 李传亮. 油藏工程原理[M]. 3版. 北京:石油工业出版社,2017.

[3] 王家宏. 实用砂岩油藏水驱开发设计分析方法[M]. 北京:石油工业出版社,2016.

[4] Tarek Ahmed. Reservoir Engineering Handbook. Cambridge:Gulf Professional Publishing,2001.

[5] 陈涛平. 石油工程[M]. 2版. 北京:石油工业出版社,2012.

[6] 刘德华,唐洪俊. 油藏工程基础[M]. 2版. 北京:石油工业出版社,2011.

[7] 陈元千. 预测油气田产量的Weibull模型[J]. 新疆石油地质,1995,16(3):250-255.

[8] 童宪章. 天然水驱和人工注水油藏的统计规律探讨[J]. 石油勘探与开发,1978,4(6):38-64.

[9] 陈元千. 一种新型水驱曲线关系式的推导及应用[J]. 石油学报,1993,14(2):65-73.